# Target Costing

# MECHANICAL ENGINEERING

A Series of Textbooks and Reference Books

*Founding Editor*

## L. L. Faulkner

*Columbus Division, Battelle Memorial Institute
and Department of Mechanical Engineering
The Ohio State University
Columbus, Ohio*

1. *Spring Designer's Handbook*, Harold Carlson
2. *Computer-Aided Graphics and Design*, Daniel L. Ryan
3. *Lubrication Fundamentals*, J. George Wills
4. *Solar Engineering for Domestic Buildings*, William A. Himmelman
5. *Applied Engineering Mechanics: Statics and Dynamics*, G. Boothroyd and C. Poli
6. *Centrifugal Pump Clinic*, Igor J. Karassik
7. *Computer-Aided Kinetics for Machine Design*, Daniel L. Ryan
8. *Plastics Products Design Handbook, Part A: Materials and Components; Part B: Processes and Design for Processes*, edited by Edward Miller
9. *Turbomachinery: Basic Theory and Applications*, Earl Logan, Jr.
10. *Vibrations of Shells and Plates*, Werner Soedel
11. *Flat and Corrugated Diaphragm Design Handbook*, Mario Di Giovanni
12. *Practical Stress Analysis in Engineering Design*, Alexander Blake
13. *An Introduction to the Design and Behavior of Bolted Joints*, John H. Bickford
14. *Optimal Engineering Design: Principles and Applications*, James N. Siddall
15. *Spring Manufacturing Handbook*, Harold Carlson
16. *Industrial Noise Control: Fundamentals and Applications*, edited by Lewis H. Bell
17. *Gears and Their Vibration: A Basic Approach to Understanding Gear Noise*, J. Derek Smith
18. *Chains for Power Transmission and Material Handling: Design and Applications Handbook*, American Chain Association
19. *Corrosion and Corrosion Protection Handbook*, edited by Philip A. Schweitzer
20. *Gear Drive Systems: Design and Application*, Peter Lynwander
21. *Controlling In-Plant Airborne Contaminants: Systems Design and Calculations*, John D. Constance
22. *CAD/CAM Systems Planning and Implementation*, Charles S. Knox
23. *Probabilistic Engineering Design: Principles and Applications*, James N. Siddall
24. *Traction Drives: Selection and Application*, Frederick W. Heilich III and Eugene E. Shube
25. *Finite Element Methods: An Introduction*, Ronald L. Huston and Chris E. Passerello
26. *Mechanical Fastening of Plastics: An Engineering Handbook*, Brayton Lincoln, Kenneth J. Gomes, and James F. Braden
27. *Lubrication in Practice: Second Edition*, edited by W. S. Robertson
28. *Principles of Automated Drafting*, Daniel L. Ryan
29. *Practical Seal Design*, edited by Leonard J. Martini
30. *Engineering Documentation for CAD/CAM Applications*, Charles S. Knox
31. *Design Dimensioning with Computer Graphics Applications*, Jerome C. Lange
32. *Mechanism Analysis: Simplified Graphical and Analytical Techniques*, Lyndon O. Barton
33. *CAD/CAM Systems: Justification, Implementation, Productivity Measurement*, Edward J. Preston, George W. Crawford, and Mark E. Coticchia
34. *Steam Plant Calculations Manual*, V. Ganapathy
35. *Design Assurance for Engineers and Managers*, John A. Burgess

36. *Heat Transfer Fluids and Systems for Process and Energy Applications*, Jasbir Singh
37. *Potential Flows: Computer Graphic Solutions*, Robert H. Kirchhoff
38. *Computer-Aided Graphics and Design: Second Edition*, Daniel L. Ryan
39. *Electronically Controlled Proportional Valves: Selection and Application*, Michael J. Tonyan, edited by Tobi Goldoftas
40. *Pressure Gauge Handbook*, AMETEK, U.S. Gauge Division, edited by Philip W. Harland
41. *Fabric Filtration for Combustion Sources: Fundamentals and Basic Technology*, R. P. Donovan
42. *Design of Mechanical Joints*, Alexander Blake
43. *CAD/CAM Dictionary*, Edward J. Preston, George W. Crawford, and Mark E. Coticchia
44. *Machinery Adhesives for Locking, Retaining, and Sealing*, Girard S. Haviland
45. *Couplings and Joints: Design, Selection, and Application*, Jon R. Mancuso
46. *Shaft Alignment Handbook*, John Piotrowski
47. *BASIC Programs for Steam Plant Engineers: Boilers, Combustion, Fluid Flow, and Heat Transfer*, V. Ganapathy
48. *Solving Mechanical Design Problems with Computer Graphics*, Jerome C. Lange
49. *Plastics Gearing: Selection and Application*, Clifford E. Adams
50. *Clutches and Brakes: Design and Selection*, William C. Orthwein
51. *Transducers in Mechanical and Electronic Design*, Harry L. Trietley
52. *Metallurgical Applications of Shock-Wave and High-Strain-Rate Phenomena*, edited by Lawrence E. Murr, Karl P. Staudhammer, and Marc A. Meyers
53. *Magnesium Products Design*, Robert S. Busk
54. *How to Integrate CAD/CAM Systems: Management and Technology*, William D. Engelke
55. *Cam Design and Manufacture: Second Edition; with cam design software for the IBM PC and compatibles, disk included*, Preben W. Jensen
56. *Solid-State AC Motor Controls: Selection and Application*, Sylvester Campbell
57. *Fundamentals of Robotics*, David D. Ardayfio
58. *Belt Selection and Application for Engineers*, edited by Wallace D. Erickson
59. *Developing Three-Dimensional CAD Software with the IBM PC*, C. Stan Wei
60. *Organizing Data for CIM Applications*, Charles S. Knox, with contributions by Thomas C. Boos, Ross S. Culverhouse, and Paul F. Muchnicki
61. *Computer-Aided Simulation in Railway Dynamics*, by Rao V. Dukkipati and Joseph R. Amyot
62. *Fiber-Reinforced Composites: Materials, Manufacturing, and Design*, P. K. Mallick
63. *Photoelectric Sensors and Controls: Selection and Application*, Scott M. Juds
64. *Finite Element Analysis with Personal Computers*, Edward R. Champion, Jr., and J. Michael Ensminger
65. *Ultrasonics: Fundamentals, Technology, Applications: Second Edition, Revised and Expanded*, Dale Ensminger
66. *Applied Finite Element Modeling: Practical Problem Solving for Engineers*, Jeffrey M. Steele
67. *Measurement and Instrumentation in Engineering: Principles and Basic Laboratory Experiments*, Francis S. Tse and Ivan E. Morse
68. *Centrifugal Pump Clinic: Second Edition, Revised and Expanded*, Igor J. Karassik
69. *Practical Stress Analysis in Engineering Design: Second Edition, Revised and Expanded*, Alexander Blake
70. *An Introduction to the Design and Behavior of Bolted Joints: Second Edition, Revised and Expanded*, John H. Bickford
71. *High Vacuum Technology: A Practical Guide*, Marsbed H. Hablanian
72. *Pressure Sensors: Selection and Application*, Duane Tandeske
73. *Zinc Handbook: Properties, Processing, and Use in Design*, Frank Porter
74. *Thermal Fatigue of Metals*, Andrzej Weronski and Tadeusz Hejwowski
75. *Classical and Modern Mechanisms for Engineers and Inventors*, Preben W. Jensen
76. *Handbook of Electronic Package Design*, edited by Michael Pecht
77. *Shock-Wave and High-Strain-Rate Phenomena in Materials*, edited by Marc A. Meyers, Lawrence E. Murr, and Karl P. Staudhammer
78. *Industrial Refrigeration: Principles, Design and Applications*, P. C. Koelet

79. *Applied Combustion*, Eugene L. Keating
80. *Engine Oils and Automotive Lubrication*, edited by Wilfried J. Bartz
81. *Mechanism Analysis: Simplified and Graphical Techniques, Second Edition, Revised and Expanded*, Lyndon O. Barton
82. *Fundamental Fluid Mechanics for the Practicing Engineer*, James W. Murdock
83. *Fiber-Reinforced Composites: Materials, Manufacturing, and Design, Second Edition, Revised and Expanded*, P. K. Mallick
84. *Numerical Methods for Engineering Applications*, Edward R. Champion, Jr.
85. *Turbomachinery: Basic Theory and Applications, Second Edition, Revised and Expanded*, Earl Logan, Jr.
86. *Vibrations of Shells and Plates: Second Edition, Revised and Expanded*, Werner Soedel
87. *Steam Plant Calculations Manual: Second Edition, Revised and Expanded*, V. Ganapathy
88. *Industrial Noise Control: Fundamentals and Applications, Second Edition, Revised and Expanded*, Lewis H. Bell and Douglas H. Bell
89. *Finite Elements: Their Design and Performance*, Richard H. MacNeal
90. *Mechanical Properties of Polymers and Composites: Second Edition, Revised and Expanded*, Lawrence E. Nielsen and Robert F. Landel
91. *Mechanical Wear Prediction and Prevention*, Raymond G. Bayer
92. *Mechanical Power Transmission Components*, edited by David W. South and Jon R. Mancuso
93. *Handbook of Turbomachinery*, edited by Earl Logan, Jr.
94. *Engineering Documentation Control Practices and Procedures*, Ray E. Monahan
95. *Refractory Linings: Thermomechanical Design and Applications*, Charles A. Schacht
96. *Geometric Dimensioning and Tolerancing: Applications and Techniques for Use in Design, Manufacturing, and Inspection*, James D. Meadows
97. *An Introduction to the Design and Behavior of Bolted Joints: Third Edition, Revised and Expanded*, John H. Bickford
98. *Shaft Alignment Handbook: Second Edition, Revised and Expanded*, John Piotrowski
99. *Computer-Aided Design of Polymer-Matrix Composite Structures*, edited by S. V. Hoa
100. *Friction Science and Technology*, Peter J. Blau
101. *Introduction to Plastics and Composites: Mechanical Properties and Engineering Applications*, Edward Miller
102. *Practical Fracture Mechanics in Design*, Alexander Blake
103. *Pump Characteristics and Applications*, Michael W. Volk
104. *Optical Principles and Technology for Engineers*, James E. Stewart
105. *Optimizing the Shape of Mechanical Elements and Structures*, A. A. Seireg and Jorge Rodriguez
106. *Kinematics and Dynamics of Machinery*, Vladimír Stejskal and Michael Valášek
107. *Shaft Seals for Dynamic Applications*, Les Horve
108. *Reliability-Based Mechanical Design*, edited by Thomas A. Cruse
109. *Mechanical Fastening, Joining, and Assembly*, James A. Speck
110. *Turbomachinery Fluid Dynamics and Heat Transfer*, edited by Chunill Hah
111. *High-Vacuum Technology: A Practical Guide, Second Edition, Revised and Expanded*, Marsbed H. Hablanian
112. *Geometric Dimensioning and Tolerancing: Workbook and Answerbook*, James D. Meadows
113. *Handbook of Materials Selection for Engineering Applications*, edited by G. T. Murray
114. *Handbook of Thermoplastic Piping System Design*, Thomas Sixsmith and Reinhard Hanselka
115. *Practical Guide to Finite Elements: A Solid Mechanics Approach*, Steven M. Lepi
116. *Applied Computational Fluid Dynamics*, edited by Vijay K. Garg
117. *Fluid Sealing Technology*, Heinz K. Muller and Bernard S. Nau
118. *Friction and Lubrication in Mechanical Design*, A. A. Seireg
119. *Influence Functions and Matrices*, Yuri A. Melnikov
120. *Mechanical Analysis of Electronic Packaging Systems*, Stephen A. McKeown

121. *Couplings and Joints: Design, Selection, and Application, Second Edition, Revised and Expanded,* Jon R. Mancuso
122. *Thermodynamics: Processes and Applications,* Earl Logan, Jr.
123. *Gear Noise and Vibration,* J. Derek Smith
124. *Practical Fluid Mechanics for Engineering Applications,* John J. Bloomer
125. *Handbook of Hydraulic Fluid Technology,* edited by George E. Totten
126. *Heat Exchanger Design Handbook,* T. Kuppan
127. *Designing for Product Sound Quality,* Richard H. Lyon
128. *Probability Applications in Mechanical Design,* Franklin E. Fisher and Joy R. Fisher
129. *Nickel Alloys,* edited by Ulrich Heubner
130. *Rotating Machinery Vibration: Problem Analysis and Troubleshooting,* Maurice L. Adams, Jr.
131. *Formulas for Dynamic Analysis,* Ronald Huston and C. Q. Liu
132. *Handbook of Machinery Dynamics,* Lynn L. Faulkner and Earl Logan, Jr.
133. *Rapid Prototyping Technology: Selection and Application,* Ken Cooper
134. *Reciprocating Machinery Dynamics: Design and Analysis,* Abdulla S. Rangwala
135. *Maintenance Excellence: Optimizing Equipment Life-Cycle Decisions,* edited by John D. Campbell and Andrew K. S. Jardine
136. *Practical Guide to Industrial Boiler Systems,* Ralph L. Vandagriff
137. *Lubrication Fundamentals: Second Edition, Revised and Expanded,* D. M. Pirro and A. A. Wessol
138. *Mechanical Life Cycle Handbook: Good Environmental Design and Manufacturing,* edited by Mahendra S. Hundal
139. *Micromachining of Engineering Materials,* edited by Joseph McGeough
140. *Control Strategies for Dynamic Systems: Design and Implementation,* John H. Lumkes, Jr.
141. *Practical Guide to Pressure Vessel Manufacturing,* Sunil Pullarcot
142. *Nondestructive Evaluation: Theory, Techniques, and Applications,* edited by Peter J. Shull
143. *Diesel Engine Engineering: Dynamics, Design, and Control,* Andrei Makartchouk
144. *Handbook of Machine Tool Analysis,* Ioan D. Marinescu, Constantin Ispas, and Dan Boboc
145. *Implementing Concurrent Engineering in Small Companies,* Susan Carlson Skalak
146. *Practical Guide to the Packaging of Electronics: Thermal and Mechanical Design and Analysis,* Ali Jamnia
147. *Bearing Design in Machinery: Engineering Tribology and Lubrication,* Avraham Harnoy
148. *Mechanical Reliability Improvement: Probability and Statistics for Experi-mental Testing,* R. E. Little
149. *Industrial Boilers and Heat Recovery Steam Generators: Design, Applications, and Calculations,* V. Ganapathy
150. *The CAD Guidebook: A Basic Manual for Understanding and Improving Computer-Aided Design,* Stephen J. Schoonmaker
151. *Industrial Noise Control and Acoustics,* Randall F. Barron
152. *Mechanical Properties of Engineering Materials,* Wolé Soboyejo
153. *Reliability Verification, Testing, and Analysis in Engineering Design,* Gary S. Wasserman
154. *Fundamental Mechanics of Fluids: Third Edition,* I. G. Currie
155. *Intermediate Heat Transfer,* Kau-Fui Vincent Wong
156. *HVAC Water Chillers and Cooling Towers: Fundamentals, Application, and Operations,* Herbert W. Stanford III
157. *Gear Noise and Vibration: Second Edition, Revised and Expanded,* J. Derek Smith
158. *Handbook of Turbomachinery: Second Edition, Revised and Expanded,* Earl Logan, Jr., and Ramendra Roy
159. *Piping and Pipeline Engineering: Design, Construction, Maintenance, Integrity, and Repair,* George A. Antaki
160. *Turbomachinery: Design and Theory,* Rama S. R. Gorla and Aijaz Ahmed Khan

161. *Target Costing: Market-Driven Product Design,* M. Bradford Clifton, Henry M. B. Bird, Robert E. Albano, and Wesley P. Townsend
162. *Fluidized Bed Combustion,* Simeon N. Oka
163. *Theory of Dimensioning: An Introduction to Parameterizing Geometric Models,* Vijay Srinivasan

**Additional Volumes in Preparation**

*Structural Analysis of Polymeric Composite Materials,* Mark E. Tuttle

*Handbook of Pneumatic Conveying Engineering,* David Mills, Mark G. Jones, and Vijay K. Agarwal

*Handbook of Mechanical Design Based on Material Composition,* George E. Totten, Lin Xie, and Kiyoshi Funatani

*Mechanical Wear Fundamentals and Testing: Second Edition, Revised and Expanded,* Raymond G. Bayer

*Engineering Design for Wear: Second Edition, Revised and Expanded,* Raymond G. Bayer

*Clutches and Brakes: Design and Selection, Second Edition,* William C. Orthwein

*Progressing Cavity Pumps, Downhole Pumps, and Mudmotors,* Lev Nelik

*Mechanical Engineering Software*

*Spring Design with an IBM PC,* Al Dietrich

*Mechanical Design Failure Analysis: With Failure Analysis System Software for the IBM PC,* David G. Ullman

# Target Costing
## Market-Driven Product Design

**M. Bradford Clifton**
*Lucent Technologies*
*Holmdel, New Jersey, U.S.A.*

**Henry M. B. Bird**
**Robert E. Albano**
*Enterprise Systems Engineering Group*
*Princeton, New Jersey, U.S.A.*

**Wesley P. Townsend**
*Bell Laboratories, Lucent Technologies*
*Princeton, New Jersey, U.S.A.*

CRC Press is an imprint of the
Taylor & Francis Group, an informa business

Although great care has been taken to provide accurate and current information, neither the author(s) nor the publisher, nor anyone else associated with this publication, shall be liable for any loss, damage, or liability directly or indirectly caused or alleged to be caused by this book. The material contained herein is not intended to provide specific advice or recommendations for any specific situation.

Trademark notice: Product or corporate names may be trademarks or registered trademarks and are used only for identification and explanation without intent to infringe.

**Library of Congress Cataloging-in-Publication Data**
A catalog record for this book is available from the Library of Congress.

**ISBN: 0-8247-4611-2**

This book is printed on acid-free paper.

**Headquarters**
Marcel Dekker, Inc.
270 Madison Avenue, New York, NY 10016, U.S.A.
tel: 212-696-9000; fax: 212-685-4540

**Distribution and Customer Service**
Marcel Dekker, Inc.
Cimarron Road, Monticello, New York 12701, U.S.A.
tel: 800-228-1160; fax: 845-796-1772

**Eastern Hemisphere Distribution**
Marcel Dekker AG
Hutgasse 4, Postfach 812, CH-4001 Basel, Switzerland
tel: 41-61-260-6300; fax: 41-61-260-6333

**World Wide Web**
http://www.dekker.com

The publisher offers discounts on this book when ordered in bulk quantities. For more information, write to Special Sales/Professional Marketing at the headquarters address above.

**Copyright © 2004 by Marcel Dekker, Inc. All Rights Reserved.**

Neither this book nor any part may be reproduced or transmitted in any form or by any means, electronic or mechanical, including photocopying, microfilming, and recording, or by any information storage and retrieval system, without permission in writing from the publisher.

Reprinted 2010 by CRC Press
CRC Press
6000 Broken Sound Parkway, NW
Suite 300, Boca Raton, FL 33487
270 Madison Avenue
New York, NY 10016
2 Park Square, Milton Park
Abingdon, Oxon OX14 4RN, UK

To our families, and to those in industry and academia with the wisdom and courage to apply the principles and methods of target costing.

# Preface

This book is about a simple, straightforward process that can have a significant impact on the health and profitability of many businesses. Target costing doesn't require an army of specialists, large-scale software implementation, or complex management structures and procedures. It consists of mostly logical, disciplined common sense that can be embedded into a company's existing procedures and processes.

Along with our colleagues we have spent our recent professional careers applying the principles of target costing to a wide range of products, processes, and procedures in a large manufacturing company. We quickly came to learn, and fervently believe, that target costing helps to achieve these goals:

- Ensuring that products are better matched to their customers' needs
- Reducing the development cycle of a product
- Reducing the costs of products significantly
- Fostering teamwork between all internal organizations associated with conceiving, marketing, planning, developing, manufacturing, selling, distributing, and installing a product
- Engaging customers and suppliers to design the right product and to more effectively integrate the entire supply chain

We have seen first-hand the huge improvements the target costing allows. We have also heard of many similar examples from colleagues in other companies across a wide range of industries. Feeling compelled to share what we have learned with others in industry, we decided to write this book.

There is already a growing body of literature about target costing, and some very good textbooks appeared a few years ago when target costing was emerging in the United States. We felt that it was time to offer a practical book on the subject—one with greater emphasis on practical concerns (the "how to" of target costing) rather than the underlying theory. We also introduce numerous concepts and approaches that we have developed and added to our practice of target costing; we believe that these approaches expand the capabilities of target costing.

This book closely follows a training course that we have given many times within our own company and that has also been taught in a modified form at a major university. It is organized in a logical, step-by-step fashion that guides the reader through the sequence of steps that must be executed when applying target costing. Also included are the examples, checklists, and so forth that the reader would expect to find in this type of book.

But this book adds a unique element—the Exercise. The reader is strongly urged to take the time to work through the phases of the Exercise that appear at the end of each chapter. Then he or she will get some actual experience with doing the steps in the target costing process, and will see a business result at the end. For use in a classroom environment, we encourage the instructor to divide the class into teams of five to six people and let each team try to achieve the best business result. Individually or as a team, it's fun! So work through this book, and then go out and help your company achieve better business success!

*M. Bradford Clifton*
*Henry M. B. Bird*
*Robert E. Albano*
*Wesley P. Townsend*

# Acknowledgments

We wish to acknowledge those who have gone before us—the developers of the first target costing concepts and the early adopters of those concepts. We have been guided and inspired by those who have written and spoken about target costing, and we have benefited from many private conversations with its early practitioners. In this book we have scrupulously tried to acknowledge those sources and references. If we have omitted an appropriate reference, we offer our sincere apologies. Any reader who notes such an omission should rest assured that it was inadvertent; please notify us so that can be corrected.

Many of the figures and tables in the examples provided are borrowed or adapted from our many target costing projects at Lucent Technologies. The source material is contained in internal documents and publications that are not generally available. Scores of colleagues, as members of cross-functional teams, contributed significantly to the project and results. They are far too numerous to list here, but we wish to acknowledge their efforts, their collaborative spirit, and their insights. The following individuals were important pioneers who provided a favorable climate for target costing and actively promoted its use: Marc Benowitz, Jean Lous Bourdon, Cynthia Chenault, Bill Clifford, Steve Condra, Dick Donovan, Peter Fitton, Sidney Heath III, Bob Henn, Dave Lando, Jose Mejia, Arthur Meyer, Glenn Moyer, Robert Piconi, Charlotte Ramstek, Saied Seghatoleslami, Bill Sessa, and Shishir Thanawala.

Several people contributed their comments and suggestions during the preparation of this book. We wish to thank Judy Bird, Rod Clifton, Lee L'Esperance, and especially Marc Weinstein.

# Contents

*Preface*     *v*
*Acknowledgments*     *vii*

| | | |
|---|---|---|
| Chapter 1 | Introduction | 1 |
| Chapter 2 | Define the Product | 15 |
| Chapter 3 | Set the Target (Product Level) | 39 |
| Chapter 4 | Set the Target (Subsystem Level) | 55 |
| Chapter 5 | Achieve the Target | 71 |
| Chapter 6 | Maintain Competitive Costs | 89 |
| Chapter 7 | Putting Target Costing into Practice | 101 |
| Chapter 8 | Some Case Histories | 113 |
| Chapter 9 | Wrap-Up and Conclusion | 135 |
| | | |
| Module A | Create a Business with a Strategy | 137 |
| Module B | Quantify Customers' Needs | 155 |
| Module C | Determine Target Price and Cost (Product Level) | 169 |
| Module D | Determine Cost Targets (Subsystem Level) | 179 |
| Module E | Find Paths to Achieve the Targets | 191 |
| Module F | Get Financial Results | 211 |
| Module G | Optimizing Results | 219 |

*Appendix: Sample Exercise*     *239*
*Glossary*     *257*
*Index*     *263*

# Target Costing

# 1

## Introduction

**THE POTENTIAL**

Imagine that you have just introduced a new product into the marketplace. How would your Vice President respond if its cost were 10% below plan? She probably wouldn't say much because you have a good team that works hard to control cost and she expects this level of cost control from you. How would your Vice President respond if your cost at introduction was *20%* below plan? She would probably be interested and ask you how you did it, but little more would come of it. But how would your Vice President respond if your cost was *30%* below plan? She would most likely be very interested and ask how you did it so she could apply your methodology to every product in her division. This is how Target Costing gets started in some firms. Target Costing is the approach that is helping many firms achieve costs significantly lower than they have in the past. Target Costing has been shown to consistently reduce product costs by up to 20–40%, depending on your circumstances. Through the text and the Exercise, this book presents a proven approach to achieving competitive product cost applicable to a variety of industries. This quick-read book provides users practical Target Costing tools and methods that achieve significant cost savings for their firms.

But what is Target Costing? Our definition, adapted from Cooper and Slagmulder [1], is as follows:

> Target Costing is a disciplined process for determining and realizing a total cost at which a proposed product with specified functionality *must* be produced to generate the desired profitability at its anticipated selling price in the future.

There are several important points in this definition that should be emphasized:

1. A disciplined process.
2. Specified functionality.
3. Desired profit.
4. Anticipated selling price.

Target Costing is a disciplined process that uses data in a logical series of operations to determine and achieve a target cost for the product. In addition, the price and cost are for specified product functionality. The functionality is determined from understanding the needs of the customer and the willingness of the customer to pay for each function.

Another interesting aspect of Target Costing is its inherent recognition that the important variables in the process are essentially beyond the control of the design group or even the company. For example, the selling price will be determined by the marketplace, the global collection of customers, competitors, and the general economic conditions at the time the product is introduced. The desired profit is another variable that is beyond the control of the design organization. It may be set at the corporate level. It will be influenced by the expectation of the stockholders and the financial markets. And the desired profit is benchmarked against others in the same industry and against all businesses. In this complicated environment, it is the role of Target Costing to balance these external variables and develop the product at a cost that is within the constraints imposed. In short, a simple cost-plus approach is a recipe for failure, while giving the customers more than they are willing to pay for is a recipe for insolvency.

Much of what is included in Target Costing comes from the cost-management process used by Japanese firms. It has been adapted to use best practices already in place in European and American firms. Target Costing is a relatively new topic to many firms in the United States, but many of the supporting processes are already in place. Some books on Target Costing are written from the finance perspective. This book emphasizes the marketing and engineering aspects, where many of the cost-saving breakthroughs occur. Much of the material for this book is drawn from our experiences working in an electronics company in Japan, from many Target Costing applications in Lucent Technologies over a wide range of telecommunications equipment, and from our friends and colleagues who have been applying Target Costing in other companies.

## MOTIVATION

Why should there be a renewed emphasis on applying Target Costing to products? In the late 1990s, business and Wall Street were on a roll. It was very easy to be successful, so businesses and individuals prospered. Then the bubble burst. The perception is that it was only the technology

# Introduction 3

sector—especially the dot.coms—that collapsed significantly. But the contraction is much more widespread. The question today is: "**Now that times are tough, what do you need to do to be successful?**"

There are three fundamental things that companies need to do to prosper:

- Offer products that are exactly what customers want and need.
- Offer products in a timely fashion when customers want to buy.
- Sell them at competitive prices (and so keep costs low).

Target Costing is usually thought to be associated with the latter issue. But when it is properly applied in the *complete* product-development cycle, it can also have significant impact on all three issues. The points above are obvious—almost trite—and companies put plenty of energy into addressing them. So why should an enterprise need Target Costing? Because their sincere and best efforts often don't achieve the intended results.

How often have you heard the following remarks?

"We aren't price competitive in the eyes of the customer."
"We don't understand customer requirements—in terms of priorities and willingness to pay."
"We don't understand our product's *total* cost structure or our customer's total cost of ownership."
"We don't align our value proposition with that of the customer."
"Why are we not meeting our business plan for profits and market share?"
"We don't sufficiently understand competitors' offerings in terms of pricing strategies, cost structures, product strengths and weaknesses."
"We don't target-cost our products early enough in new product introduction (NPI) to gain the greatest cost savings."
"We don't leverage our supply line early enough in NPI."

## WHEN ARE COSTS SET?

Ask yourself, "At what point are costs set or locked in?" We don't mean when they are incurred—that happens when the product is in production and being sold to customers. We mean, "When are the decisions made that determine what the costs will be?" Most people assume that it's when the developers and designers make their decisions about what the product will look like, what its architecture will be, what subsystems and components will be used, and what the manufacturing process will be. In fact, much of a product's cost is locked in even before the design starts, as shown in Fig. 1.1.

This figure (adapted from Smith & Reinertsen [2]) shows that costs begin to be set at the earliest phases of product realization. By the time the product gets to the development team, more than 50% of the eventual

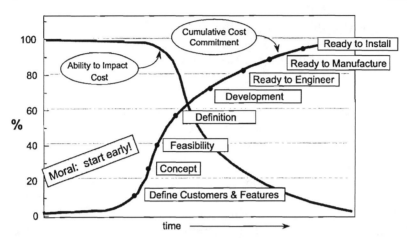

**Figure 1.1** Target Costing should be applied early in the product-realization process, when the team has the greatest ability to impact cost.

costs can already be committed or locked in. In addition, the ability to influence cost has greatly diminished by the time the product development gets to the traditional areas of cost reduction which focus on materials and manufacturing costs. For this reason, Target Costing should begin at the very beginning of the so-called "fuzzy front end" of the product-realization process. We have found a strong correlation between the cost savings obtained and how early in the product-realization cycle Target Costing was initiated.

Traditional cost-reduction efforts tend to focus on "cost of goods sold" (COGS), which includes the purchased materials and subsystems that go into a product, plus the costs to convert them into the final product (often called "labor and load"). Also, these "continuous improvement" efforts (*kaizen*, in Japanese) tend to be most effective later in the cycle, after the product is in manufacture. Later we will say more about doing Target Costing on *total* costs. Here, the important concept is that *Target Costing begins with Marketing*, and requires very good knowledge of the customers, the competitors, and the marketplace.

## THE BASIC TARGET-COSTING PROCESS

Fig. 1.2 shows the four basic steps in Target Costing. It begins with the definition of the product, carries through setting the target cost, finding ways to achieve and achieving the target, and then maintaining a competitive cost during the life cycle of the product.

The process begins at the front end of the product's realization, when the product is being conceived. The concept can arise from direct input from the marketplace (customers), from people with an intimate knowledge of the marketplace (Marketing), or as an idea about new ways

# Introduction

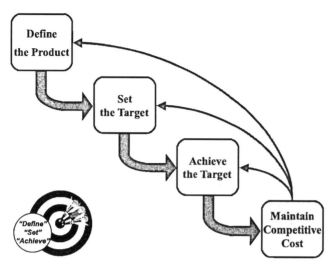

**Figure 1.2** The basic Target Costing process.

to meet a need (Research and Development), understanding that there is an unsatisfied need in the market. There is one other case where Target Costing should be applied early in the cycle—when an existing product is about to undergo a major revision, release or redesign. In all cases, *the product-definition step is strongly market-based.*

In setting the target costs, it is important to *start with price*. You must know what the customers are willing to pay for the product and its various capabilities. That often means understanding the customer's total costs of ownership (TCO). It also means knowing what the competitors are charging, and what their cost structures are. Only after determining the price—and the required profit margin necessary for business health—can you determine the target cost. It is important to state that the cost is the dependent variable, price and margin are the independent variables:

Target cost = Target price − Target margin

Achieving the target means finding ways in which the product can be designed and produced while meeting the target cost. This usually requires out-of-the-box thinking, brainstorming, the inputs and ideas of a wide range of people in the product chain—and above all, a relentless discipline to question all assumptions about every cost element in the product. This should be *done very early in the design cycle*, and it means making "cost" a design parameter as important as any other feature or specification of the product.

Once the ideas for eliminating cost have been proposed, they need to be evaluated for feasibility and effectiveness, and then they need to be *put into practice* during the detailed design phase and during production of the product. One aspect of this step is to avoid the "feature creep" phenomenon that often occurs while a product is being designed. If the firm learns that a new feature is required during development, it must come with an associated benefit and cost. To include it may require tradeoffs that are sometimes better postponed to the next product release.

Finally, once the product has been introduced at the target cost, competition continues and prices tend to decline. Therefore *continuous improvement* is necessary; this includes traditional cost-reduction efforts. But they have a relatively small effect because they are applied after most of the significant decisions that affect cost have been made.

The process is a continuous and dynamic one, as suggested by the small reverse arrows in Fig. 1.2. There should be constant checking of results at each stage, and careful feedback to and checking of assumptions in previous stages. For example, in maintaining competitive cost it is necessary to keep a careful finger on the pulse of the marketplace, checking for changes in, for example, desired features or eroding prices. These will affect "Define the Product" and "Set the Target."

Fig. 1.3 shows the four-step process again, with the fundamental questions that need to be dealt with at each step, as explained above. In the chapters that follow we will describe the details of what happens in each step.

Many might say, "This is too simple. Besides, in our company we do this anyway." Yes, at this level it looks extremely simple—but that is part of the reason for its success. It provides strong focus, and you know exactly

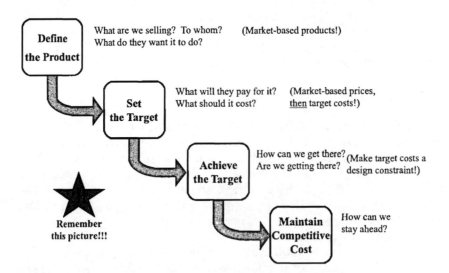

**Figure 1.3** The fundamental questions at each step in Target Costing.

# Introduction

where you are at any point in time. Moreover, it will become clear that the procedures that need to be done *within* each step create a comprehensive and systematic approach to product development that leads inexorably to achieving the right product at the right cost. It is the logic and the *discipline* of Target Costing that makes it work. Our definition of Target Costing, stated before, emphasizes this:

> Target Costing is a disciplined process for determining and realizing a total cost at which a proposed product with specified functionality *must* be produced to generate the desired profitability at its anticipated selling price in the future.

## EXAMPLE: ELECTRONICS CABINETS

A telecommunications manufacturer made equipment, and the outdoor cabinets in which its equipment could be housed. The cabinets were sometimes sold bundled with the equipment and sometimes as stand-alone products. The company found itself losing market share in both the equipment and the cabinets, and part of the cause was the high price (due to high cost) of the cabinets. They decided to redesign the product line, including the cabinets.

A team was assembled to deal with the cabinets, and Target Costing played a crucial role in this project. To regain market share and meet financial goals, the company had to offer customers what they wanted at a competitive price. Tying together elements of Target Costing such as competitive tear-downs, market analysis, and design for manufacturability, the team redesigned the cabinets and the manufacturing process to achieve a 30% cost saving over the previous designs across the product line. With its new product line, the firm's revitalized cabinets business is growing again.

The cabinet project started with a small, empowered, cross-functional core team. The core team had representatives from Manufacturing, Design, Sales, Product Management, Market Management, and R&D. Depending on the nature of the product, such teams may also include representatives from Customer Service or Field Service. The team reviewed recent changes in the market, baseline costs, competitive analysis information, new design concepts, and financial goals. The team quantified the differences in cost and features that made the cabinets uncompetitive. This review allowed each discipline represented to see its role in the entire enterprise.

Following this meeting, the team embarked on a Target Costing project. As a team they conducted competitive product tear-downs and shared the results across the business unit. Next they formulated a customer survey and invited customers to participate in the design of the next-generation cabinet product platform. Combining inputs from the competitive information, customer feature and price requirements, and gross margin goals, they set market-based cost targets. To brainstorm solutions to achieve the target, the team called in expertise beyond their small team and even beyond their business unit. In a day-long meeting, the situation was reviewed and then the challenge was presented to the wider cross-functional team. This group watched as the core team took each subsystem and further broke it down to the component level. Interfaces between the subsystems were well documented so everyone could see which issues extended across subsystems. At this level, the outside experts could contribute their brainstormed solutions best without regard for any perceived constraints of the cabinet business. After the meeting, the core team considered all the concepts from the brainstorming. They clarified concepts, validated assumptions, conducted experiments, obtained quotes, and made the necessary tradeoffs to agree on a path to achieve the cost targets before proceeding into the design phase. This approach allowed the core team to benefit from the wide expertise of the entire company, without a large investment of time by the outsiders.

The result one year later was a new family of cabinets that met the customer needs, beat the competition in critical price and performance metrics, and that had a cost that helped return the business to high profitability.

**EXAMPLE: TELECOMMUNICATIONS EQUIPMENT**

This story [3] is illustrated in Fig. 1.4. Another division of the same company had been first-to-market in a major new telecom business. It enjoyed considerable success for a few years, and design and engineering efforts had achieved cost reductions amounting to 14%. Then competitors began to offer products with new features and at competitive prices. It was time to design the next generation of the product family.

The design team quickly produced a version that looked as if it could be produced at much lower cost—39% below the cost-reduced first-generation product, or about half the cost of the original product. This was a significant achievement! At about this time, Target Costing was beginning its life in the company, and it was suggested that a Target Costing look at the product might be useful. Of course this meant that Target Costing was initiated later than it should be, but the team started the effort anyway.

The team went back to the "front end" of the process and examined the marketplace, the competitors, the customers' needs, and their willingness to pay. They discovered that, at the anticipated date of product availability, the price (and hence the cost) had to be an additional 31% lower than the original redesigned product! This was disheartening, and was met with some skepticism. However, an analysis of pricing trends from concurrent major bids revealed that the prices in the marketplace were indeed declining at a rate that supported the required

# Introduction

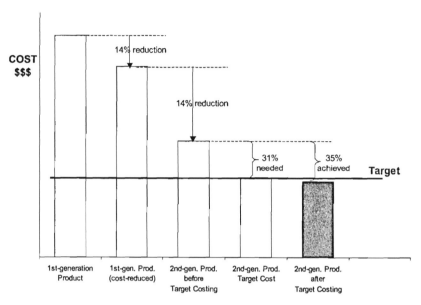

**Figure 1.4** Applying Target Costing to an existing high-technology product.

31% further reduction in cost. So the target cost was accepted as a necessary requirement.

After this, the story unfolded in a way similar to the previous example. A multidisciplinary team was formed from within the business unit, and included subject-matter experts from elsewhere in the company. Product management drove the effort, with strong support from R&D. Representatives with expertise in competitive intelligence, design, systems engineering, manufacturing, purchasing, cost analysis, and installation participated vigorously. They were united by a common vision of the customer's needs combined with the required competitive cost. After brainstorming generated over one hundred cost-reduction and feature-revision ideas, those ideas were evaluated for feasibility and cost impact. A set of implementation recommendations was made to the senior management, and adopted.

The subsequent steps of the detailed development cycle took 12–15 months, which was considered rapid compared to previous products of comparable complexity and sophistication. The main factor enabling the success was that there was a strong focus on "Achieve the Target." Through the entire 15 months, the view of projected costs did not deviate by more than 1 or 2% from the target. This meant that while the marketplace absolutely insisted that additional features be added, the development team found further ways to incorporate the needs of the evolving market and also to remove cost.

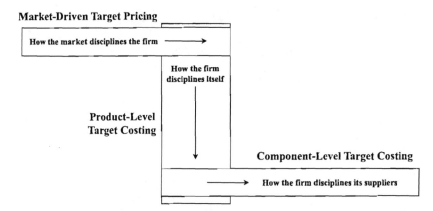

**Figure 1.5**  Discipline in the Target Costing process.

## TARGET COSTING AND DISCIPLINE

We have said that Target Costing forces a certain level of discipline on the product-realization process. Cooper [4] says that there are actually three levels of discipline that are enforced. First the marketplace disciplines the firm by imposing market-based prices. Second, the firm disciplines itself by applying Target Costing to its products. And finally, the firm disciplines its suppliers by applying Target Costing at the level of purchased materials, components, or subsystems. We will say more about these levels in subsequent chapters, but for now Fig. 1.5, adapted from Cooper, neatly illustrates the point.

## SOME WORDS ABOUT TOTAL COSTS

While most readers will be very familiar with accounting and financial concepts and terms, it is useful to provide a quick review of a few basics, in the light of our definition:

> Target Costing is a disciplined process for determining and realizing a total cost at which a proposed product with specified functionality *must* be produced to generate the desired profitability at its anticipated selling price in the future.

There are several important points to make about this definition. First, Target Costing is a disciplined and logical process, and this is the glue that holds the product-realization process together. Second, Target Costing addresses the *total* cost. There are various interpretations as to what is meant by total cost. The most common view is derived from the *pro forma* income statement. A more complete discussion of income statements can be found in any basic accounting textbook, but for this discussion see the simplified sample in Fig. 1.6 below. Revenue is the first entry and for our purposes is the revenue from sales of a product. The cost

# Introduction 11

| | | |
|---|---|---|
| Revenues | $175M | |
| Cost of Goods Sold | -$90M | |
| Gross Margin | $85M | (49%) |
| Marketing & Sales | -$18M | |
| Distribution | -$6M | |
| Research & Development | -$15M | |
| General & Administrative | -$5M | |
| Operating Income | $41M | (23%) |
| Depreciation | -$8M | |
| Taxes | -$13M | |
| Net Income | $20M | (11%) |

**Figure 1.6** Sample income statement illustrating total costs for an enterprise.

directly associated with the production of the product sold is commonly referred to as the cost of goods sold (COGS). The Gross Margin is the difference between the Revenue and COGS for the product and is one of the basic metrics of business performance. The percent gross margin is also a common metric of business performance. In this book Gross Margin is one of the drivers of target cost calculations.

There are additional entries on the income statement including Marketing and Sales, Transportation and Distribution, Research and Development, General and Administrative, Depreciation, and Tariffs and Taxes. The difference between the Gross Margin and the sum of these other costs is the Net Income. In some cases Depreciation and Taxes are excluded, and the result is called the Operating Income or Operating Margin. These costs enter into the determination of the total target costs associated with a product.

These other expenses contain costs beyond the manufacture of the product and the composite represents total costs. Since much of the R&D and other expenses precede the COGS and revenue associated with a sale, some prefer to look at the total costs associated with a product's realization and order realization processes.

## OBJECTIVES AND BENEFITS OF TARGET COSTING

The obvious primary benefits of Target Costing are:

- Having the right products.
- Having competitive prices.
- Making a profit.

Target Costing gets its effectiveness from its market-driven approach to cost management. Sometimes cost objectives are set for teams based upon previous versions of a product, or even arbitrarily. Target Costing rigorously pursues and uses inputs from competitive analyses and customers to establish targets. Teams are more motivated when they understand that there is a rational external reason for the ambitious goals that they are being asked to achieve. In general, performance is better when aimed toward a specific goal rather than an unspecified "best effort" goal.

Target Costing is also effective because the results come from using the strengths of the entire firm, and not just the development and engineering team. In today's rapidly changing world, it is nearly impossible to introduce a product with a small isolated design team. There are legal, technical, marketing, and business issues that must all be satisfied. As illustrated in the examples above, the results come from using the knowledge and expertise of the many different functions in a firm and even beyond the firm.

Therefore, it is not surprising that there are subsidiary benefits. Horvath [5] suggests that Target Costing can:

- Orient the company toward the marketplace.
- Strategically link R&D with the needs of customers.
- Facilitate and support cost management in the early design phases of a product.
- Enable firms to actively manage costs by providing cost targets that can be periodically reviewed.
- Help the company meet its financial goals, thus preserving the livelihood of its customers, employees, and shareholders.
- Motivate employees through the use of market-based requirements rather than abstract company goals.

**ABOUT THIS BOOK**

Although Target Costing may be seen as new, you and your colleagues already know many of the tools it uses. But you have yet to integrate them into your standard practices. This book will help you take this step. It will help you "Design the Right Product at the Right Cost, and Delight Your Customers while Making a Profit." Teams that have used this Target Costing method have achieved product costs 20–40% lower than they would have without it, and the benefits extend beyond product cost. We believe that you will achieve similar results.

The rest of this book will take you, step by step, through the Target Costing process. We want to give the reader the ability to go out and apply it in his or her business. Therefore, we emphasize the practical, "how to" aspects of the discipline. We do indicate some of the theoretical or academic aspects of the topic, but these are not emphasized. We provide

# Introduction 13

examples and checklists, and indicate tools that can be used. We also discuss some of the organizational and "soft" issues that are often encountered when applying Target Costing for the first few times in an enterprise.

We also provide some very specific and detailed information on certain tools and techniques. This is set aside in sections labeled "For Practitioners." These sections give the practitioner who wishes to pursue detailed analysis and/or understand the inner workings of the techniques more details than might be desired by those who wish only to use the methods. The general reader may skip over those sections without any loss of ability to understand or apply the general concepts of Target Costing.

The concepts are illustrated with an Exercise involving the development of a hypothetical product. The Exercise is based on a course developed and presented many times by the authors for the Lucent Technologies Product Management Education program. You will be taken through the process of bringing the new product into the marketplace, and you will determine a financial result. In this process you will set a strategy, get marketplace information, set targets, design the product, find paths to achieve the targets, and introduce it into the marketplace at prices of your choosing. You will then see how your product fares against versatile competitors. You can still learn much about Target Costing if you merely read the text and skim the Exercise, but you will be far more knowledgeable and better prepared to apply Target Costing if you gain some experience by doing the Exercise. Finally, building off the exercise, we present a profit-optimization method that uses the Target Costing method to provide a framework for solving the reader's design problems.

It is intended that you alternate between the text and the Exercise. At the end of each chapter you will be directed to go to the next module of the Exercise.

> You should now start the Exercise.
> Please turn to Module A, page 137.

## REFERENCES

1. Lucent Technologies' version of definition proposed by Cooper, Robin; Slagmulder, Regine, *Target Costing and Value Engineering*, Portland, Oregon: Productivity Press, 1997.
2. Adapted from Smith Preston G. & Reinertsen, Donald G., *Developing Products in Half the Time*, New York: Van Nostrand Reinhold, 1995, p. 225
3. This and many other examples are taken from the authors' work at Lucent Technologies.

4. Cooper, Robin, *When Lean Enterprises Collide: Competing Through Confrontation*, Boston: Harvard Business School Press, 1995.
5. Adapted from—Horvath, Peter, Universität Stuttgart, "Target Cost Management in German Companies—a Case Study of Siemens," The First Annual International Congress on Target Costing Conference Proceedings, 1997.

# 2

## Define the Product

**WHERE ARE WE IN THE PROCESS?**

We are now about to begin detailed discussion of the Target Costing process itself. In this chapter, we will focus on the step "Define the Product." Relative to our high-level view of the five-step process, we are at the first step, as shown in Fig. 2.1. As will seen, this chapter could as well have been entitled "Quantify Customer's Needs." This is because most of the activities are focused on the specific features that the product will contain. This information is critical since the feature set will have a dominant impact on the eventual cost of the product.

We have emphasized that Target Costing is market-driven. It starts with a clear view of the marketplace—the customers and the competitors. It is also important to have a corporate or business-unit strategy, and to be sure that the product planning is consistent with the strategy. These three things—customers, competitors, and strategy—are the essential inputs at the "Define the Product" step in the Target Costing process, as shown in Fig. 2.2. The figure is based on Ansari's definitions [1]:

- *Market Research* provides quantitative information about the needs and wants of customers.
- *Competitive Analysis* determines what competitors' products are currently available to the target customers, how the customers evaluate these other products, and how competitors might react to your company's new product introductions.
- *Product Planning* involves analyzing market and competitor information to decide what particular customer segments to concentrate on.

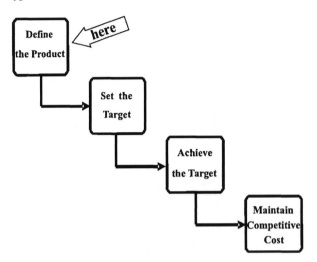

**Figure 2.1** We are at the first step in the process.

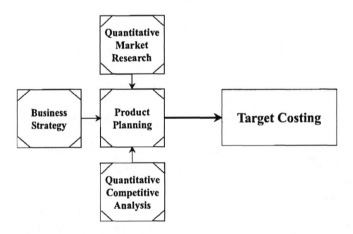

**Figure 2.2** Fundamental inputs to Target Costing.

## PRODUCT SUCCESS AND FAILURE

At this point, it is instructive to think about why some products are successful and why some fail. IBM [2] did a study of its products over a period of many years, and rank-ordered the principal reasons for failure. Results are shown in Fig. 2.3. (The sum is more than 100% because some failures had multiple causes.) It is interesting to note that "Misunderstood Wants and Needs" was the biggest cause of product failures. The authors have seen similar evidence at other companies. When the customer requirements are not clear, the product development team can go in many directions and often does. This leads to product development

### Define the Product 17

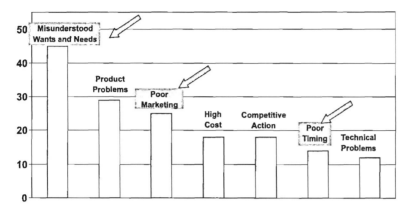

**Figure 2.3** Some reasons why products fail (from IBM study[2]).

delays because time is spent developing alternative approaches to meet a product need or repeated revisions of the same solution as changes are made to match the latest view of customer needs. In the worst cases, misunderstanding customer wants and needs leads to a much less desirable product. Sometimes this is manifested as "Poor Marketing" and "Poor Timing," which were also significant contributors to product failures according to the IBM study. This data shows the importance of good product definition. Therefore, in this book we will emphasize understanding customer wants and needs. This can be done directly or indirectly. We will start with some indirect methods.

## COMPETITOR KNOWLEDGE

Looking at your competitors is one way to gather information about your customers. Their products show what the customers are willing to buy. There may be other factors too, like proximity, relationships, or technology alignment, so it is important to put your knowledge of your competitors in context. Start with a qualitative comparison between your company and your competitors. Fig. 2.4 shows a simple table to organize your competitor knowledge.

With this context, you can turn to the more quantitative data. The collection of this information will enhance your qualitative view and likely give you new insights. Although some of the information is hard to come by, much of it is readily available. A few examples are included in Fig. 2.5.

The most direct information is on similar products. It can come from reverse engineering—that is to say, purchasing, dismantling, and studying the competitors' products. Consumer products are readily available at retail outlets. Other items may be more difficult to obtain, but in most cases it can be done legally and ethically. The outlay may be considerable, but the cost of ignorance can be much higher. Once in hand,

|  | Strategy | Value Proposition | Strengths | Weaknesses |
|---|---|---|---|---|
| Your Company | | | | |
| Competitor A | | | | |
| Competitor B | | | | |
| Others | | | | |

**Figure 2.4** Qualitative comparison of your company to your competitors.

| General Information | Product-Specific |
|---|---|
| annual reports of public companies | product catalogs and brochures |
| press releases | trade show presentations |
| job advertisements | advertisements |
| investment brokers | public bids |
| www.hoovers.com | lost bids |
| web sites | |
| consultants' and analysts' market reports | |
| Government agencies (US Department of Commerce, patent filings, SEC filings) | |

**Figure 2.5** Some sources of information about competitors.

you can analyze them to determine what components and suppliers were used, estimate costs, hypothesize about technology trends, see what customers were willing to buy. Although this is a good source of quantitative information, remember that if you were able to obtain it, it is probably old and the competitor is likely working on the next-generation solution.

## CUSTOMER KNOWLEDGE

This is *the* critical area, where success or failure begins. Most companies sincerely believe that they have a good knowledge of their customers and their needs. After all, a company may have been in business for many years, it may be a market leader, it may have long-standing and comfortable relationships with its customers, and it may spend large amounts of money on "marketing" and "market research." And yet, how many times are companies surprised when their customers switch to a competitor, or find a different way to do something that now doesn't require the original product or service? And how many companies are surprised when they launch a new product, with all the latest features and capabilities, into the marketplace and it has very poor sales?

## Define the Product

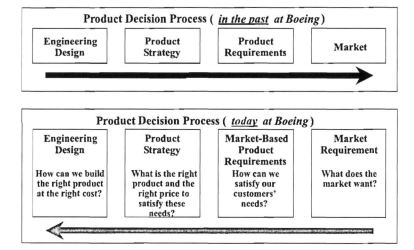

**Figure 2.6** Boeing Aircraft's product-decision—past and present (from Ref. 3).

There is a simple dictum in marketing that claims customers want to get everything and vendors want to provide nothing. Actual products are somewhere in the middle. Target Costing helps define where the middle ground is. It is essential to understand the customers' willingness to pay for the features in a product. The customers' willingness to pay for a feature or function determines exactly what is needed. Offering more features will make the customer very happy, but at increased cost to the company resulting in lower profit margins. The following example from Boeing Aircraft [3] demonstrates the point. There are many others and the reader will no doubt have several experiences as well.

For many years, Boeing Aircraft had a virtual world-wide monopoly on commercial aircraft. As one of their product managers said, "Our engineers would come up with a technical marvel, and we'd push it out into the marketplace." But then Airbus Industries appeared on the scene with an increasingly advanced line of increasingly competitive aircraft, well designed for the target markets. At one point recently, Airbus actually exceeded Boeing in market share. Boeing began to take a more market-based approach, letting the requirements be dictated by the marketplace, shown in Fig. 2.6.

The Boeing 777 is a case in point. The primary design criterion was, "Denver to Honolulu, fully loaded, on a summer's day." That defined the size (mid-size), the range, and the fact that it had to have enough fuel and lift to take off at 5000 ft altitude in the thin hot summer air. But to get to the next, more detailed level of customer wants and needs, Boeing realized that it had to have more focus. In the past Boeing had offered every option that its customers conceived. This was causing extreme complexity, logistical nightmares, and congestion in the production facilities. For example, the paint shop had 127 shades of white paint (Air France white, British Airways white, Continental Airlines white, etc.). Airlines could decide whether they wanted the air-conditioning ducts to be routed through the cargo compartments on the starboard or the port side. That single choice affected the placement of 2500 other parts in the aircraft. So when Boeing started work on the 777, they decided that they would design the aircraft based on what the top 6–8 customers wanted. That would represent the "basic" aircraft and options. Any other options desired by other airlines would come with a steep price tag.

Moreover, Boeing went out of its way to get detailed input from those top customers. Pilots, flight attendants, mechanics, and even passengers were invited to participate in focus groups and design reviews to be sure that their needs were reflected. The plane that emerged was designed to closely match the wants and needs of the principal customers, and even the end customers (the flying public).

Boeing also found that limiting the basic features to the critical decision-making features did not reduce sales. In addition (and this is a major point in Target Costing), Boeing found by that making the less critical features optional and with an added price, many customer were unwilling to pay for these extra features. The lesson is clear and supports the general rule stated above, "customers want everything." The fact that a customer desires a feature is not sufficient justification. The customer must be willing to pay for the feature. By not understanding the customers' willingness to pay, you will not be as profitable or you could even lose money. More will be said on quantifying the customers' willingness to pay and value of features later in this chapter and in the following chapter.

**MARKET-FEATURE TABLE**

We have found that one of the most powerful ways to get a clear fix on what features a new product should have is to create a one-page "market-feature" table. It is a simple matrix of the markets in columns and a hierarchy of features in the rows, as in Fig. 2.7. We emphasize that this table should be *one page*. Therefore it includes only the high-level features, decision-making functions, and characteristics of the product. It is not a detailed "requirements document" or "engineering specification."

The "natural" markets are listed across the top of the table (three in the figure), in decreasing order of size. The so-called natural markets may be regional (e.g., New England, South-East, Central, etc., or Europe, Pacific Rim, United States, etc.). They may be type of customer (e.g., commuter airlines, international carriers, etc., for aircraft; or budget shopper, teen shopper, luxury shopper, etc., for consumer products; or hobbyist, professional, etc., for photographic equipment). They may be by

# Define the Product

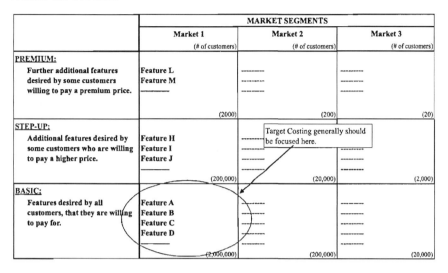

**Figure 2.7** The concept of the market-feature table.

type of industry (e.g., insurance, financial services, personnel placement, etc.). It depends on your product, your industry, and your customers.

The table is divided into three rows: from the bottom they are Basic, Step-Up, and Premium. *Basic* features are those desired by all customers in the given natural market segment, and for which all customers are willing to pay. In the figure we have filled in only the first column: here the customers in Market 1 regard features A, B, C, D, etc., as Basic. Note that other market segments may or may not regard the same features as Basic. Also, the Basic product may not be complete; it is just the version of the product that has all the features that *all* the customers in a segment want and are willing to pay for. (For example, all refrigerators require a door. Budget-minded customers would regard a plain door as "Basic," while upscale customers might consider a door with built-in ice-maker as "Basic.")

*Step-Up* features can be thought of as optional features. They are features that some—but not necessarily all—customers in a market segment want and are willing to pay for. They carry an extra price; if one customer wants a Step-Up feature, he or she will pay an additional amount for it. But the customer who doesn't want it can forgo it, and avoid the extra price. It is important to recognize that some customers will insist that these features be available, but may use them in only some applications.

*Premium* features are further features desired by some customers so much that they will pay a relatively high price for them. Again, if other customers aren't interested, they can ignore these features.

There is an extremely important corollary to this: if you put a Step-Up or Premium feature into the Basic product, *you're forcing all customers to pay for it.* In the marketplace this is often greeted by a decline in sales. Product managers have sometimes lamented, "I don't know what's wrong with my customers! We just came out with a product that has everything in it that they ever asked for. And now our sales are falling." When asked about what the customers are actually buying, the answer is, "They started buying from my competitor down the street." And when asked about the competitor's product, the answer is, "Oh, it's just a cheapie, plain-vanilla product. It doesn't do half the things that our product does." In such cases, the failure wasn't in finding out what customers thought they wanted—it was in finding out if they were willing to pay for what they wanted.

We also mentioned that each box in the market-feature table should include an indication of the size of the market for Basic, Step-Up, and Premium. The "size" could be expressed in number of customers (as in the example in the figure), or the number of units sold, or in expected revenues. By looking at the market-feature table one can quickly get an idea of what features are really important and should be incorporated into the product as soon as possible. The table can also suggest which features ought to be delayed to later releases, or even dropped altogether. In some cases it turns out that the Basic product has the same set of features across all market segments. In that happy event, you really have only one "global" market to deal with, which can be satisfied by a single Basic product, and all other features are just the merged Step-Up and Premium sets.

We have found that it is best to direct the Target Costing efforts at the Basic product for the largest one or two market segments. That will cover the majority of your addressable market, and it is where you will face the stiffest competition. That is where you *must* be price- and cost-competitive. If you are not, your competitors will grab the dominant market share, capture most of the profits, and will gain experience to prepare them for the next round of product development.

## AN EXAMPLE: ELECTRIC CAN OPENER

As an example, consider a simple hypothetical case built around the humble electric can opener. (This example has its origins in one involving an electric pencil sharpener that can be found in Ansari, et al. [1, pp. 129–131]. It worked well in our classes, so we adapted and built upon it.) It is a familiar product, and certainly simpler than the wrist video phone that you are dealing with in the Exercise. Pretend that this company has been selling high-quality, high-priced can openers into the luxury market. This company is recognized for the fine design and craftsmanship of its products. Its products are customized and prestigious, and customers who desire them are willing to pay a very high price that returns a high profit. The problem is that the number of such customers is diminishing.

**Define the Product**  23

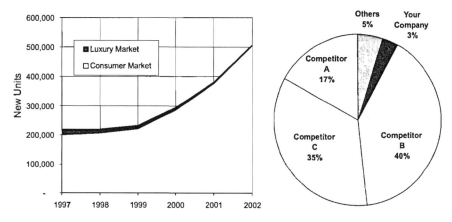

**Figure 2.8** (Hypothetical) total addressable market, and market shares in 2000, for electric can openers.

The competitors already have well-established channels to reach the broader base of customers that this company has targeted for its growth opportunities. Competitor A has an extensive presence in direct sales markets, including internet orders. Competitor B dominates overseas markets. Competitor C dominates the U.S. retail markets. Fig. 2.8 shows the (hypothetical) addressable market, and the distribution of market shares in 2000.

Some obvious points, from the market data shown in Fig. 2.8, are:

1. The total market is growing from 220,000 to 500,000 over the six-year span of the market data.
2. The Consumer segment has all the growth.
3. The Luxury segment is shrinking.
4. Our hypothetical company is the smallest of the providers.

Now let us see what the market-feature table looks like for electric can openers. There are various ways of segmenting the market—by region, sales channel, or type of customer. Fig. 2.9 indicates that there are really only two natural market segments—the huge Consumer market and the small (and shrinking) Luxury market. The (hypothetical) number of units sold is for the year 2000.

It is clear that Consumers want a simple, serviceable product that just "does the job" of opening a can and collecting the lid conveniently. They also want the product to be safe—hidden cutting blades, no sharp edges left on the can, and ability to collect the lid safely. Some consumers (but not many) will pay a little extra to get a primary color other than black, while a few will pay a little more for gold or silver. On the other hand, the Luxury segment wants an opener that does the job quickly, collects the lid, and looks beautiful. Some will pay more for cordless

|  | MARKET SEGMENTS | |
|---|---|---|
|  | CONSUMER SEGMENT (units/yr.) | LUXURY SEGMENT (units/yr.) |
| **Premium:** <br> Further additional features desired by some customers willing to pay a premium price. | Gold or Silver <br><br> 5,000 | Designer Patterns <br> Signed Limited Edition <br> 500 |
| **Step-Up:** <br> Additional features desired by some customers willing to pay a higher price. | Red, Green or Blue <br><br> 20,000 | Long Battery Life <br> Limited Edition <br> 1,500 |
| **Basic:** <br> Features desired by all customers in the segment, that they are willing to pay for. | Opens Can <br> Collects Lid Safely <br> Black <br> 260,000 | Opens Can Fast <br> Collects Lid Safely <br> Beautiful Look <br> 10,000 |

Target-cost this product first (the variant that will sell the most units).

**Figure 2.9** The market-feature table for the electric can opener.

operation and "limited editions," while a very few people will go for designer patterns (perhaps to match the wallpaper) and the even more prestigious "signed limited edition." It is also clear that to survive in the electric can opener business, this company must be competitive with a Basic product for the Consumer market.

## QUANTIFYING NEEDS

So far, our discussion has been mostly qualitative. You now have the concept of the market-feature table, and it can work wonders in clarifying your view of your proposed product and how it relates to the customer segments. But you ought to be asking, "Yes, but what is the relative importance—to the customers—of these features or attributes?" One way to learn is to simply go out and ask them, "Tell me what are the 6–8 most important fundamental functions or characteristics of the product. Then take 100 points and distribute them across those characteristics according to their relative importance." For the electric can opener, there are really only four fundamental functions or characteristics: opening speed, convenience and safety, appearance, and price. A survey of the customers might give the results in Fig. 2.10. Note that the relative importance of each characteristic is different for each of the segments—the Consumer segment cares most about price, while appearance is the main factor for the Luxury segment. This approach provides some quantitative measure of feature requirements. However, it does not provide how much more a customer in the luxury market will pay for a desirable characteristic, like appearance. We have found Conjoint Analysis an effective method for quantifying customer needs and wants. When used fully, subtle differences between customers or even between individuals in a customer

Define the Product                                                                 25

| CAN OPENER CUSTOMER VALUE MATRIX |||
|---|---|---|
| Feature or Characteristic | Consumer Segment | Luxury Segment |
| Opening Speed | 30% | 35% |
| Convenience & Safety | 15% | 20% |
| Appearance | 5% | 35% |
| Price | 50% | 10% |
| Total: | 100% | 100% |

Figure 2.10 Customer value matrix shows relative importance of key product features or characteristics for a can opener.

company can be understood. These insights are essential for designing the right product at the right cost.

## CONJOINT ANALYSIS

Conjoint Analysis is an effective way to get the customer value marrix and much more. It is a quantitative method that identifies customers' wants and needs—and their willingness to pay. It requires direct involvement by customers, because they must answer a series of questions. Basically, the method asks the customer to either rank possibilities or make a series of paired-comparison decisions. Further refinement, where customers express the strength of the preferences, provides insights into the significance of the differences. The paired questions are similar to the tradeoffs customers make in the marketplace. For example [4], if you were designing passenger automobiles, one of the questions might be, "Which do you prefer: (a) a red car for $21,000, or (b) a yellow car for $18,000? Indicate the strength of your preference: strongly prefer, mildly prefer, slightly prefer, or no preference." This can be presented concisely as shown in Fig. 2.11.

Responses to a series of simple tradeoff questions like this from an appropriate sample of customers can give a quantitative view of the market that can be used to decide what to make. In the case of an automobile, such questions would also cover other attributes such as body style, horsepower, transmission, front- or rear-wheel drive, and price. (Throughout this book, we use *attribute* to mean the broad category and *level* to describe the possibilities for that attribute. For example, *color* is an attribute; *red* and *yellow* are levels of color.) The set of answers is then analyzed to infer the "utility" to the customer of each attribute, and then each level of those attributes. A high "buyer utility" corresponds to high value to the customer. That is to say, you can quantify how much the customer prefers particular "levels" or variations of each attribute.

**Select your preference from the following scale:**

**Figure 2.11**  A typical tradeoff question in a Conjoint Analysis survey.

Occasionally, we will digress from the general flow of the text and provide deeper coverage of a particular topic. These deep-dives are for Target Costing practitioners who need to understand why something works or want to know the mathematical foundation behind a particular tool. These sections will be denoted as **For Practitioners** and separated from the text. Other readers may skip over these sections without losing the flow.

---

**For Practitioners: Determining Buyer Utilities from Answers to Tradeoff Questions.**  To illustrate how responses to tradeoff questions are converted to buyer utilities, consider the scenario of selecting a rental car. Assume that for a given class of vehicle, all attributes are the same, but you are given a choice of color, either red or blue, and a choice of audio system, either with tape player or with CD player. When converted to a Conjoint Analysis survey, respondents would be asked to either rank the possible options or respond to paired comparisons from which rank and degree are derived. To keep it simple, we will rank the four possibilities as follows:

1. Red with Tape Player
2. Red with CD Player
3. Blue with Tape Player
4. Blue with CD Player

From this simple input, an expression for preferences can be derived. First, to simplify, convert the ranks to a score by assigning the highest score to the first-ranked combination, and the lowest score to the lowest-ranked combination. For convenience, we show this as the table in Fig. 2.12. This is the input from the customer.

From this, several useful calculations are possible. They are shown in Fig. 2.13. First, sum the values for each level. In this case, the sum for red is 7, blue is 3, tape is 6, and CD is 4. This is a quantitative measure of the respondents' preferences and for this simple example can be considered the buyer utility. In many cases, you will want to associate a price with the different levels to get a measure of the customer's willingness to pay.

---

# Define the Product

|  |  | Audio | |
|---|---|---|---|
|  |  | Tape | CD |
| Color | red | 4 | 3 |
| Color | blue | 2 | 1 |

**Figure 2.12** Customer input for rental car color and audio system.

Let us return to the can opener example for a discussion on how to interpret Conjoint Analysis results. Pretend that a Conjoint Analysis survey was done for the whole market and then sorted into results for the Consumer and Luxury segments. The results are shown in Fig. 2.14. Note that "utility" is in dimensionless units.

The charts begin to tell you a *lot* about customer wants and needs, and their willingness to pay. A quick glance reveals that respondents from these two market segments have very different views on what is important, and how important. In particular, the steep price curve for the Consumer market shows that those customers are far more price-sensitive than the Luxury segment customers. For them, above all, it must be inexpensive. On the other hand, the Luxury buyers want it to be fast and beautiful.

Now analyze the results for opening speed for both segments, as shown in Fig. 2.15. This attribute has five "levels": 2, 4, 6, 8, and 10 seconds. We see that the Consumer segment is willing to tolerate a wider range of opening times. The Luxury segment wants fast can opening, and interest falls off rapidly for longer opening times.

Even though the Consumer market buyer utilities are higher for all levels but one, it is clear that the Luxury buyers are very interested in a fast opener, while the consumers don't care as much. The steep slope of the luxury market curves tells you that the Luxury market respondents make a big distinction between the levels. In this range, the Consumer market respondents make very little distinction between 2, 4, and 6 second opening times, but above 6 seconds they also exhibit a similar steep decline.

**For Practitioners: Interpreting Buyer Utilities.** To interpret buyer utilities, first analyze the importance of the attributes, and then consider the levels within each attribute. The way to analyze attributes is to compare the *spread* in buyer utilities (difference between the maximum and minimum values). The spread is a measure of an attribute's importance. If there is a relatively large difference between the maximum and minimum values, it means that this attribute is important. That is, the customers care enough about it to distinguish a difference between the levels. If, however, there is not much difference

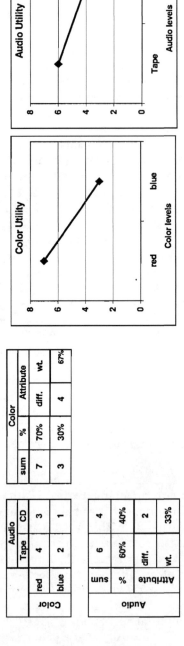

**Figure 2.13** Buyer utilities and other calculations for two rental car attributes.

# Define the Product

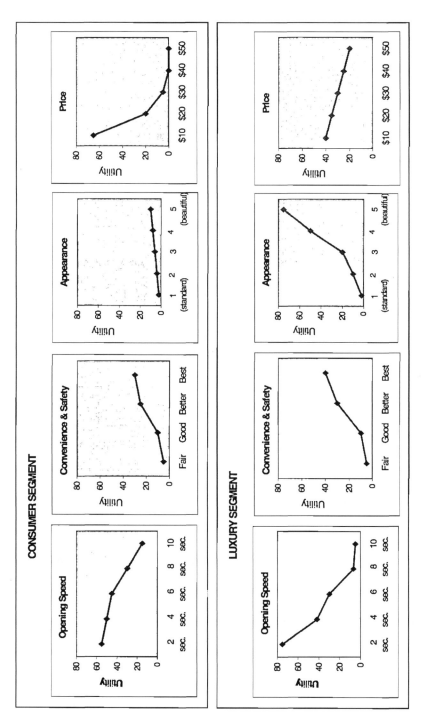

**Figure 2.14** Conjoint Analysis buyer utilities for electric can openers in two market sements.

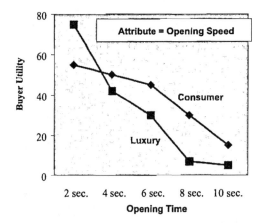

**Figure 2.15** Utilities in both segments for the speed attribute.

from one level and the next within the attribute, it means that customers do not distinguish a significant difference and so the attribute is less important. Fig. 2.16 shows how to calculate the relative importance of all the attributes for the Consumer segment, with and without price as one of the attributes. You will see later in Module C that there are reasons for separating out price.

The greatest value for Conjoint Analysis utilities is in the insights they provide on the desirability of a particular product configuration. This is especially useful when designing a new product. To illustrate, consider the low-cost can opener design in Fig. 2.17. Assume it has the following characteristics or "levels" of the attributes:

speed = 10 seconds,
convenience and safety = low,
appearance = standard, and
price = $10.

In other words, you had to sacrifice most of the attributes to achieve a rock-bottom price. How would such a product appeal to the two market segments?

Intuition tells you that the high utilities for this design in the consumer market make it more attractive to the consumer market. But be careful. To do this, you need the importance weights calculated in Fig. 2.17 to properly calculate the total buyer utility for the design. For the consumer market, the weighted total is calculated as follows:

$$P_{\text{consumer|lowcost}} = 15 * 29\% + 5 * 18\% + 2 * 6\% + 55 * 47\% = 31.3$$

# Define the Product

## Determine weighting from quantitative customer data.

Can opener example (Consumer Segment):

### Calculation of Importance Weighting

1. Determine differences from the range of each utility:
   e.g.: 55 − 15 = 40

2. Add all the differences:
   e.g.: 40 + 25 + 8 + 65 = 138

3. Determine the fraction of each attribute to the total:
   e.g.: 40/138 = 29%

4. Determine the fraction of each attribute without using the Price Utility:
   e.g.: 40/(40 + 25 + 8) = 55%

### CAN OPENER – CONSUMER SEGMENT

| Product Attributes | Utility | Difference (max−min) | Importance Weighting (all attributes) | Importance Weighting (excl. Price) |
|---|---|---|---|---|
| **Opening Time:** | | | | |
| 2 sec. | 55 | | | |
| 4 sec. | 50 | | | |
| 6 sec. | 45 | 40 | 29% | 55% |
| 8 sec. | 30 | | | |
| 10 sec. | 15 | | | |
| **Convenience & Safety:** | | | | |
| Poor | 5 | | | |
| Fair | 10 | 25 | 18% | 34% |
| Good | 25 | | | |
| High | 30 | | | |
| **Appearance:** | | | | |
| 1 (standard) | 2 | | | |
| 2 | 4 | | | |
| 3 | 6 | 8 | 6% | 11% |
| 4 | 8 | | | |
| 5 (beautiful) | 10 | | | |
| **Price:** | | | | |
| $10 | 65 | | | |
| $20 | 20 | | | |
| $30 | 5 | 65 | 47% | |
| $40 | 0 | | | |
| $50 | 0 | | | |
| **Total:** | | 138 | 100% | 100% |

### CAN OPENER – LUXURY SEGMENT

| Product Attributes | Utility | Difference (max−min) | Importance Weighting (all attributes) | Importance Weighting (excl. Price) |
|---|---|---|---|---|
| **Opening Time:** | | | | |
| 2 sec. | 75 | | | |
| 4 sec. | 42 | | | |
| 6 sec. | 30 | 70 | 35% | 39% |
| 8 sec. | 7 | | | |
| 10 sec. | 5 | | | |
| **Convenience & Safety:** | | | | |
| Poor | 5 | | | |
| Fair | 10 | 35 | 18% | 20% |
| Good | 30 | | | |
| High | 40 | | | |
| **Appearance:** | | | | |
| 1 (standard) | 2 | | | |
| 2 | 10 | | | |
| 3 | 20 | 73 | 37% | 41% |
| 4 | 50 | | | |
| 5 (beautiful) | 75 | | | |
| **Price:** | | | | |
| $10 | 40 | | | |
| $20 | 35 | | | |
| $30 | 30 | 20 | 10% | |
| $40 | 25 | | | |
| $50 | 20 | | | |
| **Total:** | | 198 | 100% | 100% |

**Figure 2.16**  Relative importance calculation of speed, convenience and safety, appearance, and price.

**Low-cost Design**

| Attribute | Level | Consumer utility | Consumer weight | Luxury utility | Luxury weight |
|---|---|---|---|---|---|
| Opening Speed | 10 sec. | 15 | 29% | 5 | 35% |
| Convenience & Safety | Fair | 5 | 18% | 5 | 18% |
| Appearance | 1 =Standard | 2 | 6% | 2 | 37% |
| Price | $10 | 55 | 47% | 40 | 10% |
| TOTAL | | 77 | 100% | 52 | 100% |

**Figure 2.17** The relative attractiveness of a low-cost can opener in the Consumer and Luxury markets.

---

**For Practitioners: Calculating Total Utilities.** You will see later that it is convenient to use matrix multiplication notation to calculate weighted totals. For a row matrix and a column matrix it takes up the same space, but as more rows or columns are added, the matrix notation simplifies the expressions. The following weighted calculation is a sum product:

$$15 * 29\% + 5 * 18\% + 2 * 6\% + 55 * 47\% = 31.3$$

It can be expressed as a multiplication of Individual Utilities and Importance Weight matrices. So, for the Consumer Market we have:

$$\underset{\text{Individual Utilities}}{(15 \ \ 5 \ \ 2 \ \ 55)} \underset{\text{Importance Weight}}{\begin{Bmatrix} 29\% \\ 18\% \\ 6\% \\ 47\% \end{Bmatrix}} = \underset{\text{Weighted total utility}}{31.3}$$

and for the Luxury market we have:

$$(5 \ \ 5 \ \ 2 \ \ 40) \begin{Bmatrix} 35\% \\ 18\% \\ 37\% \\ 10\% \end{Bmatrix} = 7.4$$

---

Clearly, this design appeals much more to the Consumer market than to the Luxury market customers. Depending on the firm's strategy, this may or may not be the right product to develop. With the buyer

## Define the Product

**Better Design**

| Attribute | Level | Consumer utility | Consumer weight | Luxury utility | Luxury weight |
|---|---|---|---|---|---|
| Opening Speed | 4 sec. | 50 | 29% | 42 | 35% |
| Convenience & Safety | Good | 10 | 18% | 10 | 18% |
| Appearance | 4= Attractive | 8 | 6% | 50 | 37% |
| Price | $20 | 55 | 47% | 35 | 10% |
| TOTAL | | 123 | 100% | 137 | 100% |

**Figure 2.18** The relative attractiveness of a better design can opener in the consumer and luxury markets.

utilities, the team is better informed about this critical decision. If the firm chooses this design, it must accept that it will have few customers, if any, from the Luxury market segment. A presence in both markets may be necessary to maintain market share. Looking at the Luxury market's weights for speed and appearance, this design will need enhancements in these areas. Now imagine a better design with the following attributes was also under consideration:

speed = 4 seconds,

convenience and safety = good,

appearance = attractive, and

price = $20

This is a better product, but it costs more and is therefore priced higher.

How would that product appeal? Although a simple sum of the utilities often shows which design is better, the reader needs to be careful, since this assumes that weights are the same for each attribute in the two market segments. In this example, merely summing will yield the wrong result. The weighted utilities calculated in Fig. 2.18 show that the better design is more appealing in the Consumer market and much more appealing in the Luxury market than the low-cost design. Apparently, customers would be willing to pay more for these enhancements.

The next question is, how much is appropriate? The answer depends on the situation. If you merely have two designs and need to pick one, as in this simple case, the insight offered by this comparison may be sufficient. It highlights the price sensitivity and trade-offs between key attributes and price for each possibility. In another situation, you may not have a specific design, but want to find the one that maximizes the weighted utility for one or both market segments based on the available attribute data. If there are only a few attributes, you could calculate and compare the possibilities in a spreadsheet or simple optimization program. In this example, the most appealing solution is the one that maximizes the Consumer market segment. This approach does not consider the cost of obtaining the solution. We will address cost and solution optimization in later chapters and in the modules.

## APPLYING CONJOINT ANALYSIS

So far, we have discussed the merits of Conjoint Analysis and introduced how to use and interpret the buyer utilities. Collecting the customer input is a significant undertaking. In applying Conjoint Analysis in complex problems, you will want to use a commercial Conjoint Analysis software package. Commercial packages do the number crunching and provide a convenient user interface. We have found the Adaptive Conjoint Analysis Tool (ACA$^©$) software and procedure provided by Sawtooth Software very useful [4]. Readers may access the website at the following URL: http://sawtoothsoftware.com. At this website, you may also evaluate the demo version of the software for your own application.

One starts by preparing a scenario to be presented to prospective customers. The scenario should describe the possibility of the new product. In the scenario, include the key attributes and levels that influence important decisions for the product. These should be the ones that define major development choices or are not clear due to diverse marketing input.

Entering them in the ACA$^©$ software is straightforward. The software generates an electronic questionnaire that can be stored on a floppy disk, downloaded from a website, or distributed through email. When interviewing prospective customers, present the scenario and ask them to provide their input about the product in the electronic survey. Then ask the respondent to put the floppy in his or her PC to run the program. The program asks them a series of questions to rank acceptable choices, assess the importance of differences among the choices, and then—based on the respondent's replies—the software calculates buyer utilities. These initial responses provide a coarse estimate of the buyer utilities. Then the program presents the comparison questions that are used to fine-tune the initial buyer utilities and determine the configuration the respondent is most likely to buy. At the end of the software "interview" (which takes only 15–20 minutes to complete), you have a complete set of buyer utilities for that respondent.

For most teams, this is more than they ever had before, and it is very useful. When combined with those of other respondents, it can model the market segments and influence design tradeoffs as was illustrated with the can opener example. Outside of the ACA$^©$ software, be sure to ask for demographic information about the respondent (e.g., name, address, type of customer, role he or she plays in the purchase decision, etc.) This is used later when sorting the data since, depending on their perspective, different customers will have different things that are important to them. For example, a vice-president of a semiconductor company will have a different view of a lithographic exposure machine than a clean-room

## Define the Product

---

**Before Customer Surveys**

1. Develop scenario to present purchase decision to customer.
2. Develop paper and computer questionnaires.
    Paper survey captures qualitative information
    - Includes demographic information (e.g. position, location)
    - Captures miscellaneous preferences (e.g. standards)
    - Business directions (maybe even forecast)
    - Allows open-ended comments
    Computer-driven Conjoint Analysis survey captures quantitative information
    - Calculates buyer utilities for key attributes in purchase decisions
    - Adapts to each respondent
3. Test surveys with your team.
4. Refine questionnaires. Collect Sales Team inputs and refine again.

---

**During Customer Surveys**

1. Present hypothetical scenario to each customer's purchase decision makers
2. Collect responses to paper survey
3. Collect responses to Sawtooth Software's ACA© survey
    Respondents narrow down preferences in up to 10 categories
    - Eliminate unacceptable attributes (only if more than 5 levels)
    - Rank levels within each attribute (all but price)
    - Choose most likely levels (only if more than 5 presented)
    - Rate impact of differences within remaining levels
    Respondents fine-tune preferences
    - Select preferred features from 2 groupings (30 paired comparisons)
    - Estimate likelihood to buy a particular combination of features (6 descriptions)

---

**After Customer Surveys**

1. Merge demographic data for sorting
2. Calculate buyer utilities
    - by company
    - with weighting by position in company, historical sales market size, etc.
3. Model scenarios
    - with external factors
    - with weighting by position in company, historical sales market size, etc.
    - by market segment

---

**Figure 2.19** Steps in the process for applying Conjoint Analysis.

engineer, and you need to know this when analyzing the replies from multiple customers. The checklists in Fig. 2.19 summarize the steps. For more on designing a Conjoint Analysis survey, consult the user's manual that comes with the software [4].

### Some Caveats

When applied, Conjoint Analysis is a powerful tool for quantifying customer preferences and is invaluable for guiding design teams. That said, there are some caveats you should be aware of (Fig. 2.20). Fortunately, there are ways to deal with each one.

| Challenge | Suggestion |
|---|---|
| Respondents' answers today may not represent the company's purchase tomorrow because situations change. | Conduct periodic surveys so your view of the market is up-to-date and stay abreast of competitors' solutions. |
| Some respondents have difficulty keeping many things in mind at the same time | Drop those with poor correlations between early responses and later responses. |
| If many attributes look the same to respondents (i.e. longevity, quality, reliability), their combined effect will outweigh price. | Test your survey on colleagues and your firm's account representatives to see how respond. Iterate and refine your survey as necessary. |
| Relative importance should be calculated using individual averages; not average utilities. | Use commercial tools to keep track of the many calculations. |
| Marketing incentives or other factors were not considered in the survey. | Aggregate and model the effect of other external factors as an additional attribute. (In simulations, be sure to calibrate their effect with known results before using with new sets of customer data.) |
| Market simulation is not a market predictor. It is only valid for relative comparisons. It usually makes many assumptions about competitors' solutions. | Consider the algorithm presented in Module G of this book that finds an optimum solution based on market and cost data when attributes of competitors' solutions are not known. |

**Figure 2.20**   Caveats when applying Conjoint Analysis.


**A WORD ON PROCESS**

With respect to "Define the Product," we show the inputs and outputs, and some of the steps that need to be taken, in Fig. 2.21. When you "Define the Product" you must understand the customer's wants and needs (and how well the competitors are already satisfying them), and then you must decide on the features and characteristics that your product must have. Organizationally, we found it best to set up a small, cross-functional steering team to drive this part of the effort. The team should be led by the Product Manager, and could include representatives from Market Research, Sales, Competitive Intelligence, Research, Product Development, Engineering, and Manufacturing. The team starts with the general market information discussed earlier, then begins to focus on the specifics of the product in question.

**SUMMARY**

It is essential to have a clear view of customers' wants and needs. To the greatest extent possible, they should be quantitative. Some other key points are:

- Form a cross-functional team early, especially to bring together the two critical inputs to Target Costing, customer requirements and competitive analysis.

**Define the Product**  37

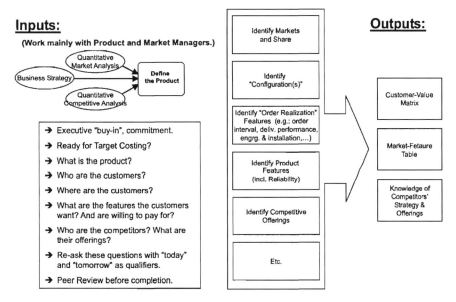

**Figure 2.21** The process associated with "Define the Product."

- Target Costing should focus on the "Basic" product that has only those features that most customers need and are willing to pay for.
- The "Basic" feature set should focus on the most significant portion of the total market.
- Conjoint Analysis is a valuable quantitative tool for measuring customer needs and willingness to pay for product features.

> You should now do the next phase of the Exercise.
> Please turn to Module B, page 155.

### REFERENCES

1. Ansari, S. L., Bell, J. E., et al., *Target Costing: The Next Frontier in Strategic Cost Management*, Consortium for Advanced Manufacturing International (CAM-I), Chicago: Irwin, 1997 (most recently listed McGraw-Hill as publisher).
2. Adiano, Cindy, IBM Consulting Group, "Using the Voice of the Customer to Improve Your Business," First Annual International

Congress on Target Costing, Conference Proceedings, 1997. (Figure was not included in proceedings, but received as a handout.)
3. Hallin, Keith, Boeing Commercial Airplane Group, "Market Target Costing and the Boeing 777 Program," First Annual International Congress on Target Costing, Conference Proceedings, 1997, p. 17.
4. Sawtooth Software, *Adaptive Conjoint Analysis User's Manual*.

# 3

## Set the Target (Product Level)

**WHERE ARE WE IN THE PROCESS?**

By now you are probably wondering, "We're into the third chapter in this book about Target Costing. When are we going to start talking about cost?" The answer is, "Soon—but not quite yet." We have discussed the first step in Target Costing: "Define the Product." The second step is "Set the Target." Notice we have not said "Set the Target Cost"; we simply said "Set the Target." The readers of fine print will ask, "How many targets need to be set?" The correct answer is, "There are many targets that will be set."

First we must discuss *price*. In this chapter, we will focus on the step "Set the Target." Relative to our high-level view of the four-step process, we are at the second step, as shown in Fig. 3.1. This chapter could have been called "Figure out the Target Price, *then* the Target Cost." In practice, this is often done in parallel with the "Definine the Product" step. This approach is consistent with the definition of Target Costing introduced in Chapter 1. Remember, the definition says "at a specified price." So the challenge in this chapter is to determine the target price before we work on the target cost. Fortunately there are useful techniques that will assist us in defining the range of our product's price.

**CONSTANT DOLLARS**

To keep things simple, in this book we assume "constant dollars," that is, that there is no inflation and the time value of money is constant. In practice, if one so desires, factors can be applied to account for inflation or deflation.

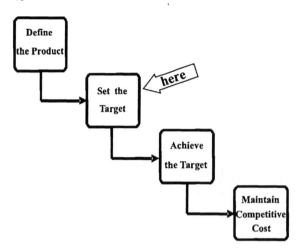

**Figure 3.1** We are at the second step in the process.

## PRICE

You first need to establish what price the customers will be willing to pay for the product, not only at the time you introduce the product into the marketplace, but also for several years into the future. This takes into account the fact that advances in technology, improvements in business efficiencies, and (above all) competition conspire to depress prices as time goes by. As Toshiro Shimoyama, Chairman and CEO of Olympus Optical Co., Ltd., has said: "Competition has become a treadmill of exhaustion from which there appears to be no escape."

Remember, in a free market the price is controlled by the customer, who always has the choice to buy or not to buy. Basic economics has demonstrated that high prices and margins attract competition, and therefore prices tend to come down as the number of units sold increases. In many companies, price is often determined merely by looking at the product, determining its actual direct costs, then adding a markup to cover overheads and provide a profit. While this "bottoms-up" approach may have worked for monopolies, or where there was little effective competition, the world has changed. Such cases are increasingly rare in today's competitive global marketplace. It is much better to use a "top-down" approach, which recognizes the realities of competitive market dynamics. Target Costing is market-based. The price that it is believed customers will pay is the first independent term in the equation:

Target cost = Target price − Target margin

We have observed that companies have much less control over their prices than they think. The two communities that actually set a company's prices are (1) its customers, and (2) its competitors.

# Set the Target (Product Level)

Clearly there are many cases where the cost-plus approach has worked and substantial profits were and probably still are being made. A sustainable competitive advantage is hard to maintain. If your firm has a good thing going, it is very likely a competitor is emerging to reap some of those benefits, too. Serendipity is not a solid foundation on which to base your company's future. Fortunately, there are quantitative techniques that can be used to help predict what customers will pay for a product before the product reaches the marketplace. Among these is the concept of the experience curve.

## EXPERIENCE CURVES

We have found that experience curves are one of the most useful ways to determine future prices and costs. They were first noted during the early 1940s in the construction of airplanes [1] (see Fig. 3.2). It was observed that the number of labor hours required to assemble an airplane decreased significantly as more and more aircraft of the same type were made:

- The second aircraft took 20% less time than the first.
- The fourth took 20% less time than the second, and so on.

The amount of labor required for the $N$th airplane $L_N$ could be expressed as a power law:

$$L_N = L_1 N^{-a} \qquad (1)$$

where $L_1$ is the number of labor hours for the first airplane and $a$ is a constant. Calculate $a$ as follows:

$$a = \log_{10}(P)/\log_{10}(N) \qquad (2)$$

where $P$ is the ratio of the $N$th time to the first one. Conveniently choosing the labor hours for the second unit,

$$a = \log_{10}(9.6/12)/\log_{10}(2) = 0.3219 \qquad (3)$$

| Unit Number | Labor (Hrs.) |
|---|---|
| 1 | 12.0 |
| 2 | 9.6 |
| 4 | 7.7 |
| 8 | 6.1 |

**Figure 3.2** Average labor hours per pound of airframe weight for World War II.

for this type of aircraft. This rate of learning is known as an "80% slope" because the ratio of the second to the first is 80%. That is to say, the number of labor hours decreases by 20% for every factor of 2 increase in cumulative experience. Combining Eqs. (1) and (2), we can write an alternative expression for the labor as a function of the %, $P$. Substituting $N = 2$ and the 80% decline for every doubling of experience, $P$, we get

$$L_N = L_1 N^{3.3219\log(P/100)} \tag{4}$$

This same relationship was also observed for fighter and transport aircraft and the trends have continued. This linear relationship between improvement and cumulative volume when plotted on a log-log scale is true in many other areas, besides labor hours to make an aircraft. When referring to the same thing each time, it is called a learning curve. When applied to experience on several different types of units, we call it an experience curve. We have observed this experience relationship in product characteristics like speed, size, and weight.

Prices are also observed to follow the same behavior, which is very useful for target costing. We find that there is a steady decrease in the price of a product for every doubling of the industry's cumulative output. If one plots the historical prices as a function of cumulative output on log-log paper, there will be scatter in the actual data points. But in general they will be fairly closely clustered—competition tends to depress them, while requirements to cover costs and achieve viable profit margins tend to elevate them. The data points tend to cluster along a declining straight line.

One of the most celebrated examples of this phenomenon is the famous "Moore's Law," an empirical relationship first discussed by Gordon Moore of Intel. Its essence is that the price of the basic function, cents per memory bit of DRAMs (Dynamic Random Access Memory), historically has declined in price by 70% for each doubling of cumulative amount of memory since the early 1960s. Fig. 3.3 shows the price of DRAM memory (expressed in price/bit) as a function of the cumulative volume in DRAM memory shipped (expressed in total bits) [2]. Note that the horizontal axis is *not* a time axis; it is a "cum-volume" axis that increases with time, but not uniformly. The dates at which the volumes and prices occurred are shown along the line. Semiconductor memory has maintained a 70% slope experience curve for three decades and over 10 quadrillion bits, with prices per bit of memory decreasing by five orders of magnitude. This dramatic micro-electronic trend, known as Moore's Law, is the technology driver for almost all the electronics industry today. Other integrated circuits, such as ASICs and DSPs, show the same type of behavior. Using curves like this, one can predict future prices of DRAMs, ASICs, DSPs, and microprocessors with a high degree of confidence.

We emphasize that you must choose the axes with some care. In the DRAM example, the vital functional element being purchased is not the

## Set the Target (Product Level)

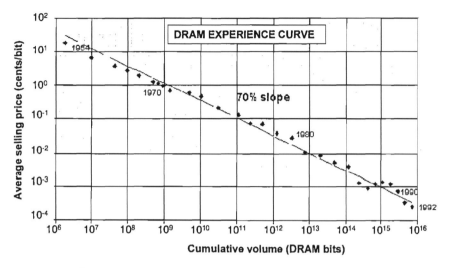

**Figure 3.3** Semiconductor DRAM experience curve.

packaged memory chips, but capacity of the memory chips as measured in bits of memory. Customers are really buying and using the ability to store a certain amount of information. That is why the horizontal axis is in bits and the vertical axis is in price per bit. Initially memory chips contained 64 bits and cost up to $10.00 each. Technology improvements have increased the capacity of the chips, and in 2001 DRAM chips with a capacity of 128 million bits (128 Mbit) cost about $20.00 retail.

For complex products, the underlying technologies that are included will be on different slopes. For example, in optical telecommunications systems the prices of the power supplies (mature) are likely to be declining much less rapidly—in both cumulative-volume and annual terms—than optical transceivers (new technology). Similarly, in aircraft the navigational instruments are likely to be on a more rapid price decline than airframe structures.

Fig. 3.4 below is an example from the world of telecommunications that demonstrates the importance of using the correct axis in an experience curve. The figure shows the prices of base stations used in cellular telephone networks. The horizontal axis is not the number of base stations or cell sites installed across the landscape, it is the cumulative number of cell-phone subscribers that exist. The service provider is buying the ability to serve a given number of subscribers; he or she is not particularly buying base stations. In buying base stations, he or she is most interested in the price that must be paid to be able to serve each subscriber. As technology improves, cell sites can individually serve greater numbers of subscribers, so the price of an individual base station may actually increase. But because far fewer base stations are required to serve a region, the average price per subscriber will decrease.

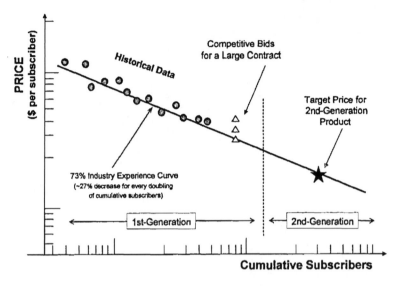

**Figure 3.4** Experience curve for cellular telephony base station prices.

Fig. 3.4 shows the price of the base station in dollars per subscriber capacity that the base station can manage (arbitrary scale) as a function of the cumulative number of cellular subscribers (also arbitrary scale) in the world. At the time the actual curve was generated in a telecommunications equipment company, that company was planning a next-generation product for deployment in about a year's time and was trying to determine what the price should be at that point in the future. The historical data of that and other companies' prices are shown as circles. At the same time there was a competitive open bid for a very large contract. The prices quoted by the three leading bidders are shown as triangles. The best bids were squarely on the experience curve defined by the historical data points. This led the company to project with high confidence the likely selling price of its next-generation product, shown by the star. As events unfolded over the next year or so, that projection proved to be correct.

One can also plot the experience curve for costs, corresponding to the price curve. If the fractional profit margins remain the same, then the two curves will be parallel on a log-log plot, as shown in the example in Fig. 3.5. It is another real case in the telecommunications industry, and shows the experience curves for both price and cost for a different kind of wireless product. As the market was just starting, Company A came out with a first-generation product at a certain price and corresponding cost. After a period of time and as cumulative sales began to increase, Company B brought out a competing product at a lower price (and cost). Company A responded some time later with its second-generation product which had an even lower price and cost. As the cumulative sales approached 70 times the early quantities, Company B brought out its second-generation

# Set the Target (Product Level)

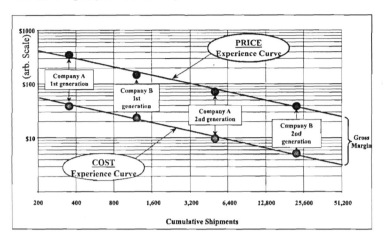

**Figure 3.5** Example of two competitors working down the experience curves for price and cost.

product at a price about a factor of 10 lower than prices prevailing as the market was getting started. Overall, both companies managed to maintain fairly constant and quite high gross margins, about 85%. Across the span of the few years covered in Fig. 3.5, the experience curves had a slope of about 70%—that is, prices (and costs) came down by about 30% for every doubling of cumulative volume.

In contrast with this particular example, however, many real product life cycles may have profit margins that change with time. But the controlling factor is the *price*. Price follows the experience curve.

Experience curves apply to a very wide range of items: airplanes, semiconductor devices, telecommunications products, steamships, elevators, electric can openers, and even eggs. Experience curves also help us realize that absolute prices tend to fall rapidly (as a function of time) early in the evolution of a product, and later fall more slowly as the product becomes widely deployed. After the market has become large, it takes a long time for the cumulative volume to double (eggs, for example).

We should also caution the reader to be sure that you are comparing products of equivalent functionality. A later-generation product with ample enhanced functions and capabilities is likely to command a higher price than a simple earlier-generation model. But if you stick to the prices of the "Basic" product (as discussed in Chapter 2) with its basic functionality and plot the prices as a function of equivalent volume, you will be on firm ground.

## OTHER WAYS TO DETERMINE TARGET PRICE

While experience curves are a compelling way to predict a future price, we have found that it is useful to make estimates by other means as well and then to "triangulate" on a likely future market price. If the date for which you want a price is in the near future, you should consider using one or more of the following methods:

### Bids and Proposals

Looking at your own company's track record of its bids, proposals, and successful *actual* selling prices. This doesn't mean catalog or list prices—it means the discounted, real prices that your customers are actually paying.

### Competitors' Prices

You can learn a lot from the trends of the prices at which your competitors are selling their products. (Again, we mean actual selling prices.) Depending on the product, this information may be available in the open marketplace, it may come from open bids and proposals, or it may come from marketplace competitive intelligence (so long as it is gathered legally and ethically).

### Reverse Engineering

This applies to costs rather than prices, but it is very useful for both validation of current costs and likely future costs when extrapolating trends. You can learn much about your competitors' technology trends, costs, and cost trends by reverse-engineering successive generations of their products. Of course, you must obtain such products by legal and ethical means.

### Bottoms-Up View

The new product may offer customers apparently new capabilities that haven't established a track record. Yet it may be possible to discern that these capabilities are combinations of existing capabilities, for which prices and experience curves have been established. In that case it may be possible to define an interpolated or composite experience curve. For example, an amphibious vehicle for sportsmen could be seen as a combination of a boat with certain more or less standard capabilities and an all-terrain vehicle with other capabilities.

### Comparable Technologies

There may not be much price information available in the marketplace, especially for a new technology. In that case, looking at the prices and price trends of similar technologies can be very suggestive. For example, the prices of new broadband services may not yet be well established. But if one looks at the trends for existing narrowband services, you may get some important clues. Moreover, often "new" technology isn't really new—in many cases it's just a new, better, and cheaper way to do the

# Set the Target (Product Level)

same thing that was being done before. In such cases, there is really a single experience curve and you must find the appropriate fundamental units to use on the axes. (For example, the experience curve for cellular base stations is close to a single straight line even when you span the successive AMPS, TDMA, CDMA and GSM technologies—so long as you use cumulative subscribers and price/subscriber for the axes.)

## Disruptive Technologies

Another way to check whether targets are reasonable is to study technology trends in other areas. Although they may look like totally different experience curves, they can describe alternative technologies that will replace the current approach. Such technology disruptions lead to drastically lower costs and/or better functionality and subsequently lower prices. Digital answering machines which have all but eliminated tape answering machines are a good example. Since disruptive technologies offer radically new capabilities, and often radically alter the cost structure, it is difficult to use them to estimate future prices. Yet, even in these cases, in the early phases of introduction, the alternative approach falls close to the established experience curves. That is because, before they reach maturity, disruptive technologies may offer only a partial set of enhancements, or they may have moderately high initial costs. In the case of digital answering machines, customers were willing to pay initially higher prices for the compactness, convenience, and consistency. Once the volume and competition grew, prices came down steeply.

## Customers' Willingness to Pay

In this case you get direct input from customers about what they would be likely to pay for your product at some point in the future. Sometimes customers even write a certain amount of price erosion into multi-year contracts. And, of course, you can use information captured by methods such as Conjoint Analysis.

## Public Information

Many government agencies and organizations are required to publish the results of all sales and purchases. This provides a rich source of competitive pricing information at a very detailed level.

## Analysts' Reports

Certain market analysis firms study industries and sell their findings for a fee. Typically they combine public information with data they collect in surveys of your firm's customers and competitors developing price trends and volume forecasts for entire industries.

## Stock Brokers

The research divisions of many stock and investment companies do in-depth analysis of industries and the major providers in the markets. Price and production estimates can often be found in these reports.

## Customers

Some customers may be willing to indicate at least the difference between your company's bid prices and those of the successful competitor, without violating any confidentiality agreements.

## TARGET COST

We are now ready (finally!) to begin discussing "cost." In setting a product's target cost, you need to know, "Which costs?" There are many costs associated with the "full stream" of selling, producing, and providing a product to customers. Let us consider manufactured products; we diagram most of these costs in Fig. 3.6.

Fig. 3.6 is based on the Income Statement used in all businesses, a graphical example of which is shown in Fig. 3.7. Notice that the single largest item in the income statement of Fig. 3.7 is the cost of goods sold (COGS), which gets most (if not all the) attention within many companies. COGS includes the cost of materials from which the product is made, and the conversion costs of transforming the materials into the finished product. Materials can be anything from "raw" (e.g., sheet steel) to purchased complex subsystems (e.g., RF amplifier modules), while the

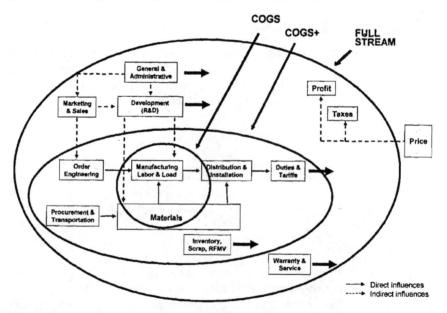

**Figure 3.6** Illustration of full-stream costs.

### Set the Target (Product Level) 49

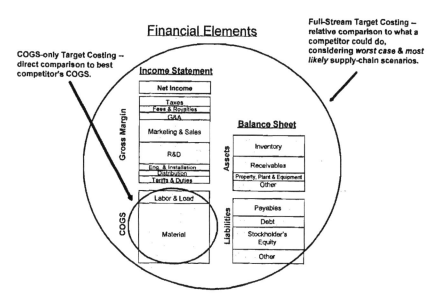

**Figure 3.7** Financial elements contained in the Income Statement and the Balance Sheet.

conversion costs are often called "labor and load," that is, the direct and indirect labor costs, plus manufacturing load or overheads (e.g., engineering, equipment maintenance, factory heating and air conditioning, the HR department, etc.). The difference between Price and COGS is usually called Gross Margin or Gross Profit, and is an important measure of a company's performance. Gross margins can vary widely from industry to industry, being higher in pharmaceuticals (lower COGS) than in telecommunications (higher COGS), for example.

COGS is seductively easy to focus on, in part because it is easy to measure and easy to relate directly to the product. It is easy because a company has a clear idea of the prices that it pays for its materials, and because factories traditionally do an excellent job of keeping track of hours worked, etc. The amount of detailed attention that is concentrated on the single issue of labor costs is remarkable. Yet, for most complex products today labor is a very small percentage of the total price, rarely exceeding 10% in the final systems assembly and testing. Other business costs, which we will discuss shortly, are often harder to measure and allocate and so they are often ignored in cost-control and Target Costing efforts.

To be sure, if you are just starting to implement Target Costing in your enterprise, it may be best to focus your first efforts on COGS only. It will make the task easier, the numbers are understood by everyone, and your positive results will be credible. In that case, you would use:

Target COGS = Target price − Target gross margin

But we encourage you to progress as quickly as possible to include more of the full-stream costs. In Fig. 3.6 we indicate what we have called "COGS+." It includes more of the supply-chain costs such as engineering the product (i.e., customizing it to meet its specific application), procurement and transportation of the materials, and delivery and installation of the product on the customer's premises. These can be a significant fraction of the total costs of complex products such as telecommunications equipment, aircraft, etc., and should be considered even as the product is being designed. For example, Dell Computer made it possible for customers to configure their computer system in real time as they place their orders on the Web. In that case, the order-engineering costs are virtually zero. But the concept worked only because Dell understood the market-feature table for its products, and designed computer products that could easily be assembled to any customer's configuration.

And when you look beyond the costs that are relatively easy to allocate to individual units, you see cost elements that are often a very significant fraction of the total price. Marketing and Sales is often the second-largest item (after Materials) in a company's costs. In many companies, Research and Development and Engineering together are the third- or even the second-largest element in the cost structure. Then there are costs such as Warranty and Service, Taxes, Depreciation, and so on. Many of these items are difficult to allocate to each specific unit of product that is sold, and so companies tend to allocate or "smear" them equally across all products, often as a percent of product revenues, material costs, or hours of labor. The only problem is that some products are probably unfairly cross-subsidizing other products, and the less profitable ones remain hidden while the stars go unrecognized. It is obvious that such subsidies can distort the apparent cost of a product and can lead to incorrect business decisions. But the subject of cost allocation is beyond the scope of this discussion.

Some of the cost elements are spread out over several years and not associated with the repetitive reproduction of what is actually sold. Product Development is a good example, and software is the poster child of this disparity. A company may spend millions to develop a software package, but only pennies to replicate the software on a CD ROM and put it in a box with the instruction booklet. (R&D is high, COGS is negligible.) This element of software makes it difficult to determine its cost in a meaningful or useful way.

Then there are financial items that do not appear on the Income Statement but on the Balance Sheet. These include assets such as Accounts Receivable, Inventories, etc., and liabilities such as Accounts Payable, Debt, etc. (see Fig. 3.7). Such items affect cash flow and the health of the company, and should be considered in Target Costing. As the product is being conceived and developed, you should be asking how the new product will affect these factors and even setting targets for some of them.

## Set the Target (Product Level)

An example of product design that can impact the balance sheet is the inventory requirements of the new product. Having many unique components to support a large selection of user configurations will add to inventory cost. A product that takes a long time to install and pass customer-acceptance testing will impact the accounts receivable portion of the balance sheet. Sales and marketing costs are a significant cost in most companies. Products that require a large amount of sales support in the form of training, configuration, etc., will increase the cost of the product. All of these will have a negative impact of the overall financial health of the company. Conversely, products with low inventories, or low sales and installation costs, will positively impact the company's financial performance.

Therefore, in your Target Costing efforts we encourage you to look at *all* factors that significantly affect your product's costs, set targets for them, and try to reduce them. Otherwise you might find yourself with your industry's lowest COGS but your competitor is taking your market share away because he has figured out how to significantly reduce his non-COGS costs. If you are looking at *everything* in the full stream, then you can use:

Target full-stream costs = Target price − Target net margin

### A WORD ON MARGINS

Whether you are doing COGS-only Target Costing, truly full-stream Target Costing, or something in between, you need to be able to subtract the appropriate margin from the price in order to arrive at the cost target. Here, you have to be guided by factors such as:

- your shareholders' expectations;
- your company's guidelines;
- your industry's norms;
- your needs to cover costs not included in your targets.

(The latter point means that, if you are doing COGS-only Target Costing, your gross margin needs to be big enough to cover all the expected non-COGS costs and leave enough net profit.) Whichever margin you use, it must be reasonable. Planning for low margin will leave you with an unhealthy business, and planning for an unrealistically high margin will lead to unattainable targets. But remember, the price is determined by the marketplace. Your desire for a high margin on a high cost of goods is a recipe for Chapter 11 (not in this book but in a Bankruptcy Court).

### A WORD ON PROCESS

The simple steps in this phase of the Target Costing process are shown in Fig. 3.8. After you "Define the Product," where you decide on the features

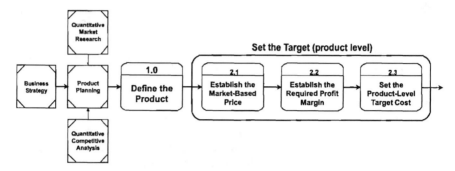

**Figure 3.8**  The basic steps in "Set the Target" (product level).

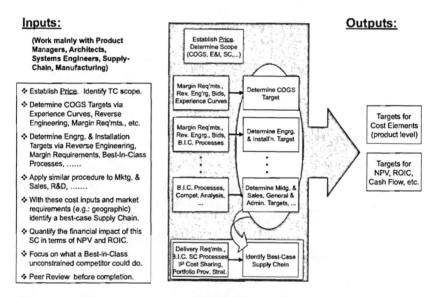

**Figure 3.9**  The process associated with "Set the Target" (at the product level).

and characteristics that your product must have, you must figure out what price your customers will pay for the product at times in the future. Then you must decide how much of the full-stream costs you want to consider and subtract a reasonable and realistic profit margin at that level of cost. What you are left with is the target cost at the level of the complete product. (Later we will discuss allocating this target cost among the subsystems that comprise the product.) Organizationally, you will probably get most of your data and information from Market Research, Competitive Intelligence, Marketing, Product Management, and Finance.

Fig. 3.9 shows this phase of the Target Costing process in more detail. Here it can be seen that, depending on what scope of costs is included, you have to determine targets for each of the major cost

## Set the Target (Product Level)

elements—COGS, Engineering and Installation, Marketing and Sales, Research and Development, Supply Chain, General and Administrative, etc. You will have to arrive at these targets by considering a variety of issues such as required profit margins, best-in-class capabilities, competitive information, reverse engineering, and so on.

## SUMMARY

In this chapter we focused on how to arrive at a target cost at the whole-product level. In summary, here are the four main points:

- The target cost is set after defining the targeted customers and knowing the features they need and are willing to pay for.
- To be competitive, the target cost *must* be derived from the market-based target price and the target profit margin.
- Which target margin you use will depend on how much of the full-stream costs you plan to try to achieve. We recommend that you go as much beyond COGS as possible.
- Experience curves are a good way for us to estimate what customers will be willing to pay in the future. Also, alternative ways exist (see page 45).

> You should now do the next phase of the Exercise.
> Please turn to Module C, page 169.

## REFERENCES

1. Asher, H., RAND Corporation, "Crawford-Strauss Study," Cost-Quantity Relationships in the Airframe Industry, 1956.
2. Lucent Technologies, internal website, DRAM experience curve. (Initial versions of this curve were presented by the Boston Consulting Group.)

# 4

# Set the Target (Subsystem Level)

**WHERE ARE WE IN THE PROCESS?**

We are still in the step "Set the Target," as shown in Fig. 4.1. In the last chapter, we discussed setting the target price and the target cost at the whole-product level. Having the price and cost for the complete product is nice, but in reality it is of little help in designing, manufacturing, and delivering a product at that target cost and price. In this chapter we will describe ways to set cost targets for the subsystems that make up the product. In most cases, products are complex enough to be made up of several distinctly different subsystems, often comprising different technologies, different design concepts and approaches, and different constraints. Also, it is often the case that different designers and engineers are responsible for different subsystems, each applying an appropriately different set of skills to their part of the problem. These individuals or subteams need specific targets so that they can properly concentrate their attentions.

**SUBDIVIDING THE TARGET COST**

Fig. 4.2 depicts the target cost divided into subsystems. The left side assumes that you have a product concept, and an initial estimate of its likely total cost, as well as estimates of the likely costs of all the subsystems that make it up. At this point some companies would simply add a profit margin and declare a price. However, you now understand that the market sets the price, and you will have determined the target cost based on the target price and the required profit margin. Suppose that the target cost of the overall product is substantially below your

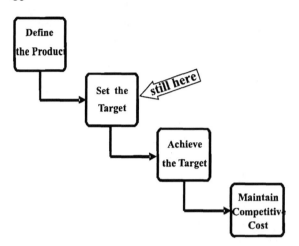

**Figure 4.1** We are still at the second step in the process.

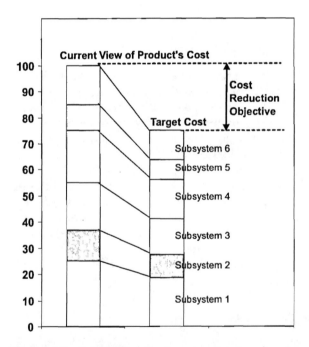

**Figure 4.2** Subdividing the target cost into subsystems.

initial estimate, as shown on the right-hand side of Fig. 4.2. Then it is reasonable to expect that you need to find ways to substantially reduce the cost of each of the subsystems. It may be tempting to apply the total required cost-reduction percentage of the whole target (say, 25%) equally to all subsystems. This is not advised for a variety of reasons. For example,

# Set the Target (Subsystem Level)

a particular subsystem may already be at its lowest practical cost and further cost reductions are not feasible. Also we have found it beneficial to make the cost targets *equally challenging* for all subsystems. That may require setting different percent cost-reduction targets for each subsystem. This chapter will describe a quantitative method based on the basic customer feature set and the customer's willingness to pay. In this approach, the target cost of each subsystem will be linked directly to customers' perceived value of the features provided by the subsystem. This quantified relationship among the features and the subsystems is a key step in determining the target cost of the product's subsystems. This is then validated with subsystem-level competitive analysis or other technology assessment.

## FUNCTIONAL TEAMS

In Chapter 2, we recommended establishing a cross-functional steering team to start your Target Costing project, to "Define the Product," and to "Set the Target" at the whole-product level. Now, as you drill down to the subsystem level, we recommend that you establish small functional teams that focus on reaching the cost target for each of the subsystems. We also recommend that there be a separate team for Manufacturing and Test—better yet, separate teams for each of Manufacturing and Test. If you are looking at more of the full-stream costs (more than just COGS), we recommend teams for other functions such as installation, transportation, etc. If you have the resources, each major cost element that you are including should have an individual cost target and a corresponding functional team composed of subject-matter experts on that topic. The subteams should be coordinated by the cross-functional steering team. Each team must know the cost target for the whole product and for each of the subsystems. The size of the team will vary based on the complexity of the subsystems and available resources, but typically, 3–6 individuals provide a good balance.

You will be tempted to populate each team only with functional experts who are very knowledgeable about the subsystem or subfunction, its functional requirements, its operation, its technology, its design, etc. We have found it extremely beneficial to "salt" these teams with individuals from outside the product or project, who have expertise in the field. If you are in a large company, it is helpful to bring in individuals from other business groups. Suppliers can bring substantial expertise to the process of meeting targets. The extent to which cost information is shared with suppliers is a question of company policy and strategy. However, if done, this approach draws on the collective expertise and experience of your entire company and its supplier base. It is impressive how many fresh ideas and concepts these "outsiders" provide. Besides fostering teamwork, it leads to better alignment of technology development and product road maps.

## A WORD ON PROCESS

We recommend that you form these teams at this point in the process, when you are starting to establish cost targets for the separate subsystems, as shown in Fig. 4.3. It is workable to have the cross-functional steering team established earlier (see Chapter 2) and set the subsystem cost targets, and not form the subteams until the end of the "Set the Target" step. But we recommend that you do it earlier, right after you have established the overall target cost. The subteams bring expertise to the process of defining the links among the features and the subsystems. In this way, the teams who are responsible for finding ways to reach the subsystem targets will have had a stake in establishing those targets. We have found that the logic inherent in the target-setting process overcomes many of the qualitative arguments that can be offered against aggressive targets.

Fig. 4.3 indicates that you must determine the value of each subsystem to the customers, make estimates of the subsystem costs (based on current knowledge), find the cost gaps, and then set the subsystem cost targets based on closing the gaps. Let us now explain this in more detail.

## VALUE-BASED PRICES AND COSTS

Like others, we have found value engineering to be a useful tool to distribute the target cost across all the subsystems or cost elements of a

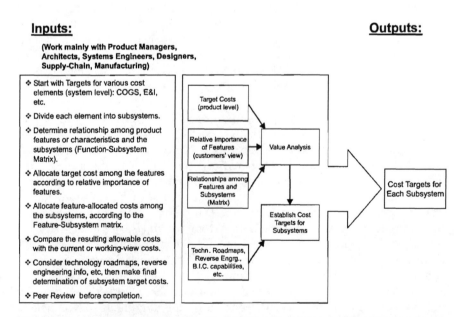

**Figure 4.3** The steps associated with "Set the Target" (at the subsystem level).

### Set the Target (Subsystem Level)

product. Value engineering is a discipline in its own right. There is considerable literature in the field; for example, Larry Shillito [1] has written and lectured extensively on the subject and provides much practical advice. We will discuss only a few specific aspects that can be applied to Target Costing.

The main principle is that the product should reflect the features that the customer values. We discussed the importance of determining the basic feature set in Chapter 2. Let us reinforce its importance. Siemens Corporation began to apply Target Costing in the mid-1990s. They started 20 projects, 2 in each of 10 divisions. Within each division they applied Target Costing to one existing product (next major release) and one new product (on the drawing boards). They averaged 30% cost reduction on the existing products, and 50% on the new products. (This reinforces Fig. 1.1!) Interestingly, two-thirds of the cost reductions came from what they called "having a market-*adequate* design" [2], as shown in Fig. 4.4. That meant that the product had *only* those features, functions, and capabilities that the customers valued. (This reinforces Chapter 2!)

There are two aspects of "market-adequate design"—meeting the functional needs and not exceeding the market price. A simple example illustrates what happens when a product is designed to meet all basic customer requirements, but for customers in distinctly different markets. In this case, a large telecommunications manufacturer found it easier for inventory control and product configuration purposes to have only one

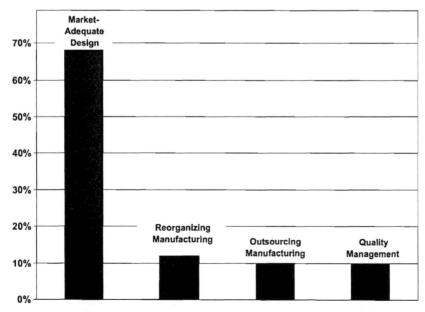

**Figure 4.4** Principal sources of cost reduction (per Siemens [2]).

outdoor enclosure for the entire world. However, the enclosure contained both an internal cooling system and an internal heating system. Some regions did not need the cooling system, while others did not need the heating system. The price of this feature-rich enclosure was 50% over the market price of the competitive products in most regions. Sales and profits for the cabinets were very poor. This product failed to satisfy the basic feature set of the distinct markets it tried to serve—customers in hot climates needed cooling, those in cold climates needed heating. The basic rule is: include only features that the customer is willing to pay for. If the customer will not pay for a feature and you still want to include it, then its cost should be zero.

It is imperative to remember that focusing on features that the customers are willing to pay for does not in any way relieve the company from providing for all the required functional elements. For illustration purposes, consider the purchase of the electric can opener. A prospective purchaser will consider the decision features such as speed, convenience and safety, and appearance. Very seldom will a customer make a selection about the cord and plug. However, no manufacturer would consider eliminating the power cord because it does not show up in the can opener purchase decision.

## QUANTIFYING THE CUSTOMERS' WANTS, NEEDS, AND WILLINGNESS TO PAY

In Chapter 2 we described ways of quantitatively determining what the customer wants, needs, and is willing to pay for. We also suggested that you determine 6 to 10 high-level aspects of the product and determine the relative importance of them to the customer (e.g., Figs. 2.8 and 2.13). But these aspects are the *customer's* view of the product, which are more functional than they are physical. How do you relate them to the *engineer's* view of the product, which is composed largely of physical subsystems? This link among the features and the subsystems is essential in setting targets. That is how value engineering can help. Let us use the humble electric can opener to illustrate the method.

## ELECTRIC CAN OPENER EXAMPLE

You will recall that the customers' view of an electric can opener is a set of features that consists of opening speed, convenience and safety, appearance, and price. In Fig. 2.8 we showed the relative importance of these to the Consumer and Luxury segments, as determined by a simple survey. In Fig. 2.15 we used more rigorous Conjoint Analysis to arrive at the relative importance to the Consumer segment of the opener's characteristics. These results, excluding price, are repeated in Fig. 4.5. For Consumers, opening speed dominates.

## Set the Target (Subsystem Level)

| CONSUMER SEGMENT (not including Price) |||
|---|---|---|
| Feature or Characteristic | Relative Importance | Value to Customers |
| Opening Speed | 55% | $11.00 |
| Convenience & Safety | 34% | $6.80 |
| Appearance | 11% | $2.20 |
| Total: | 100% | $20.00 |

**Figure 4.5** Value of features or characteristics in a customer importance matrix.

To determine the value to customers, multiply the total by the relative importance for that feature or characteristic. Let us assume that $20.00 is the price that this market segment is willing to pay for the can opener. (The buyer utilities were significantly lower—or zero—for any higher prices.) If we distribute that $20.00 according to the importance of each feature we can say, "These customers are willing to pay $11.00 for opening speed, $6.80 for convenience and safety, and only $2.20 for appearance." That is the value of those functions to those customers. The simple math is that opening speed represents 55% of the value to the customer in the total price of $20.00, or $11.00.

This is a significant determination, but if fails to provide the designers a target cost for the subsystems. From the engineer's or the designer's point of view, the electric can opener appears very different than the customers' view of the same product. The designer sees the can opener as an integration of the major subsystems:

- motor
- blades
- lid catcher
- housing.

How do these subsystems relate to the customers' perceptions of value? To determine this, we need to construct a table that links the subsystems to the features or characteristics. For complex products, this part of the Target Costing is best done by the system architects and designers, with inputs from product management. It is best to arrange the features in columns, and the subsystems in rows, as in Fig. 4.6. Then you work down each column and ask the question, "For this feature or characteristic, to what extent is it provided by each subsystem?" (The sum in each column must be 100%.) In some cases, 100% of the function is provided by a single subsystem. In other cases, a particular subsystem will not contribute to a particular function. In many cases a function will be provided by a combination of subsystems (within a column), and some subsystems will contribute to several functions (across a row). In cases like color or appearance, quantification comes naturally. For example, the electric can

|  |  | Customer Needs (Consumer Segment) |  |  |
|---|---|---|---|---|
|  |  | Opening Speed | Convenience & Safety | Appearance |
| Contribution to Needs of: | Motor | 75% | 5% | --- |
|  | Blades | 25% | 20% | 5% |
|  | Lid Catcher | --- | 75% | 15% |
|  | Housing | --- | --- | 80% |
|  | checksum | 100% | 100% | 100% |

**Figure 4.6** Electric can opener—function–subsystem matrix.

opener housing is 80% of the surface area of the product and thus contributes 80% of the appearance. For opening speed, a designer may have derived a multi-variable model of speed, blade size, and blade sharpness, but more likely a subject-matter expert made an estimate based on experience. The figures chosen create a matrix we call the function–subsystem matrix. Looking across a row, one sees where a subsystem contributes to several functions valued by customers. Fig. 4.6 shows the results for the electric can opener. Opening speed is provided by the motor and blades. Convenience and safety come from the motor, blades, and lid catcher. And the appearance is affected by the blades, lid catcher, and housing.

At this point, it is reasonable to ask, "How can a quantitative assessment be done?" In our actual experience, this is much easier that it appears. Here you must rely on the experience of the cross-functional team. With its combined expertise the team usually arrives at the feature subsystem table quickly. There may be several iterations as the features and functions are described and discussed, but the actual process is usually completed in less than a single day. You will see in larger projects that small perturbations do not make a significant difference, so precise figures are not needed.

To determine the target for each subsystem based on the value to the customer, multiply the function–subsystem matrix of Fig. 4.6 times the customer importance weight by needs matrix of Fig. 4.5. In the electric can opener example we have simplified further and are only using the Consumer segment. Therefore, calculating the value to the customer becomes a matrix multiplication as shown in Fig. 4.7. Recall that in matrix multiplication, one sums the product of the corresponding terms in the rows and columns.

These results are a measure of the value of the subsystems to the customer (in terms of price), and are shown in the right-hand column of Fig. 4.7. What we have accomplished is that we have allocated the price of the total product to the individual subsystems, and this allocation is directly linked to the customers' value of each product feature.

## Set the Target (Subsystem Level)

|  |  | Customer Needs (Consumer Segment) ||| Worth to Customer (Price) |
|---|---|---|---|---|---|
|  |  | Opening Speed | Convenience & Safety | Appearance |  |
| Customer Importance Weight → || 55% | 34% | 11% | 100% |
| Value to the Customer → || $ 11.00 | $ 6.80 | $ 2.20 | $ 20.00 |
| Contribution to Needs of: | Motor | 75% | 5% | --- | $ 8.59 |
|  | Blades | 25% | 20% | 5% | $ 4.22 |
|  | Lid Catcher | --- | 75% | 15% | $ 5.43 |
|  | Housing | --- | --- | 80% | $ 1.76 |
|  | checksum | 100% | 100% | 100% | $ 20.00 |

**Figure 4.7** What the subsystems are worth to the customer.

---

**For Practitioners: The Calculations Behind the Spreadsheets.** The total worth to the customer times the importance weight matrix yields the importance worth matrix as follows:

$$\begin{array}{ccc} \text{Total} & \text{Importance} & \text{Importance} \\ \text{Worth} & \text{Weight} & \text{Worth} \\ & \text{(by need)} & \text{(by need)} \end{array}$$

$$(\$20.00) \begin{Bmatrix} 55\% \\ 34\% \\ 11\% \end{Bmatrix} = \begin{Bmatrix} \$11.00 \\ \$6.80 \\ \$2.20 \end{Bmatrix}$$

The function–subsystem matrix times this importance worth matrix yields the subsystem worth matrix. In a spreadsheet, you will want to use the "sumproduct" function to calculate each term in the matrix as depicted by the hairpin arrow in Figs. 4.7 and 4.8:

$$\begin{array}{ccc} \text{Function–Subsystem} & \text{Importance} & \text{Subsystem} \\ \text{Matrix} & \text{Worth} & \text{Worth} \end{array}$$

$$\begin{pmatrix} 75\% & 5\% & 0\% \\ 25\% & 20\% & 5\% \\ 0\% & 75\% & 15\% \\ 0\% & 0\% & 80\% \end{pmatrix} \begin{Bmatrix} \$11.00 \\ \$6.80 \\ \$2.20 \end{Bmatrix} = \begin{Bmatrix} \$8.59 \\ \$4.22 \\ \$5.43 \\ \$1.76 \end{Bmatrix}$$

The value of the subsystems to the customer in terms of price is interesting, but we are interested in costs. Suppose we must make a certain percentage profit margin on the product. If we deduct this equally

|  | Customer Needs (Consumer Segment) |  |  | Worth to Customer (Price) | Allowable COGS (50% G.M.) | Current Working View | Value Index |
|---|---|---|---|---|---|---|---|
|  | Opening Speed | Convenience & Safety | Appearance |  |  |  |  |
| Customer Importance Weight → | 55% | 34% | 11% | 100% |  |  |  |
| Value to the Customer → | $ 11.00 | $ 6.80 | $ 2.20 | $ 20.00 | $ 10.00 | $ 11.40 |  |
| Contribution to Needs of: Motor | 75% | 5% | --- |  | $ 8.59 | $ 4.30 | $ 4.50 | 0.95 |
| Blades | 25% | 20% | 5% | $ 4.22 | $ 2.10 | $ 3.40 | 0.62 |
| Lid Catcher | --- | 75% | 15% | $ 5.43 | $ 2.72 | $ 2.10 | 1.30 |
| Housing | --- | --- | 80% | $ 1.76 | $ 0.88 | $ 1.40 | 0.63 |
| checksum | 100% | 100% | 100% | $ 20.00 | $ 10.00 | $ 11.40 |  |

**Figure 4.8** Allowable and current costs or the electric can opener.

## Set the Target (Subsystem Level)

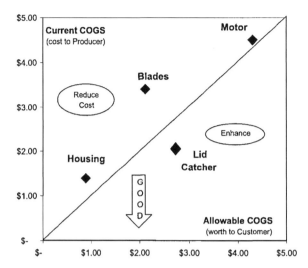

**Figure 4.9** Plot of current cost versus allowable cost.

from all the subsystem values, we can arrive at an "allowable cost" for each subsystem. In Fig. 4.8 we assumed a requirement for a 50% gross margin, and so arrived at "allowable COGS" for each subsystem. The allowable cost is that suggested by a value analysis of the target cost, based on the value of the product features and the extent to which each subsystem contributes to those features.

In the spreadsheet of Fig. 4.8, we have added a column that shows the actual current cost of the motor, blades, lid catcher, and housing for the current design. (For new products, the current working view of cost should be used instead of current costs.) You will note that some of the actual costs are higher than the allowable costs, and some are lower. This is a very common result since customers typically see only the outside of a product. Also notice that when the individual costs are summed up, the total current cost is $5.70, which is 14% higher than the $5.00 target cost (allowable cost). Therefore, some aggressive cost reduction is necessary for a successful product. The next logical question is, "How do you select the subsystems to cost reduce?"

The "value index" shown in Fig. 4.9 is simply the ratio of the allowable cost to the current or actual cost. Expressed another way:

$$\text{Value index} = \frac{\text{Worth to customer}}{\text{Cost to me}}$$

A value index of greater than 1 is good; that means that you have found a way to give the customer high value for less cost to yourself. On the other hand, if the value index is less than 1, the subsystem is costing more than the perceived value to the customer. If we plot the current view of cost versus the allowable cost for the electric can opener, we get a graph such as in Fig. 4.9. The diagonal line denotes value index = 1. Subsystems above the line (upper-left zone) represent the best opportunities for cost reduction, while those below the line (lower-right zone) are cases where you have an advantage. You can use them to subsidize some of the high-cost items, or you can add some additional features that the customers might value highly (enhance the product). Note, however, in this example the overall cost of the product exceeds its total allowable cost, so the primary effort needs to be on cost reduction. In terms of dollars, the blades and the motor represent the largest opportunities for improvement. It is often the case that there are such subsystems that are not obvious to the customer but which have a low value index.

## VALIDATION

Differences between the target as set by what the customer is willing to pay and what the available technology can provide will often be surprising to the Target Costing team. To validate these targets, it is suggested that you develop experience curves for the subsystems that contribute to your product. This is easier said than done, but goes a long way to inspire a team because it uses their own familiar data to validate the targets. When the two approaches do not agree, it leads to insights that help close gaps. Taking the electric can opener example a bit further, in Fig. 4.10 we provide hypothetical cost data for the four subsystems. In each case, available data from supplier quotes and past data on comparable products is plotted versus cumulative industry volume. Plotting on a log-log scale, the data can be approximated by a straight-line curve fit with good correlation. The dashed line projects out to the volume expected at the time for which the $10.00 total cost target was established. In this case, summing values from these subsystem level experience curves gives $9.75, thus validating the target. The current working view is $11.40, but the current total for the present time based on available components is $11.15, so there is a small current gap.

The summary in Fig. 4.11 shows that at the subsystem level, the gaps are different than what the customer-based gaps of Fig. 4.9 show. Let us examine each one. In the case of the blades, the trend line and target differ by 5%, so the target is reasonable and likely to be achieved. Both the housing and lid catcher costs are significantly higher than the target cost. This illustrates that the technology does not exist to achieve these subsystem targets. So although the customer data would imply these targets, they are unrealistic. Thus, these subsystem targets are not validated. The steering team will need to make a decision about adjusting subsystem targets. Fortunately, the motor target is well above the trend line and is easily met so a reallocation between subsystems makes each more realistic.

## Set the Target (Subsystem Level)

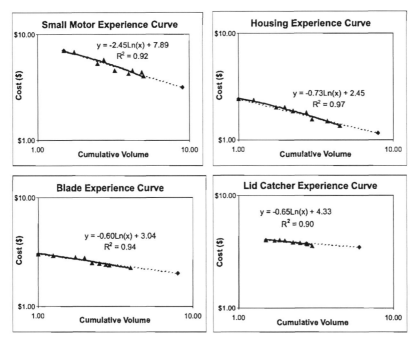

**Figure 4.10** Hypothetical experience curves for subsystems in electric can opener.

**Figure 4.11** Comparison summary of cost trend and targets for electric can opener subsystems.

## SUBSYSTEM COST TARGETS

Value analysis is a systematic way to distribute a product's overall target cost among its constituent subsystems. It leads directly to suggest allowable costs for each subsystem in a product. Should these be used as the cost targets for the subsystems? As illustrated above, the answer is, "Not necessarily." There may be constraints that make it impractical to use the allowable costs directly. For example, a particular subsystem may already be at the lowest possible cost within the bounds of technology, cost of materials, available vendor prices, manufacturing technology, etc. In that case, you may need to live with a cost for that subsystem that exceeds its target cost. But then you must remember that the target cost *is a zero-sum game*. If you leave one subsystem above its allowable cost, then you *must* set lower targets for one or more of the other subsystems. That is because the sum of all the subsystem costs *must* add up to the overall target cost, which is:

Target cost = Target price − Target margin

We have found that, unless there is a very compelling reason to do otherwise, it is best to simply accept the allowable costs suggested by value analysis and use them as the cost targets for the subsystems. It is the responsibility of the steering team to manage the process of deviation from individual subsystem target costs.

In most Target Costing projects, the current view of the cost of many or most of the subsystems is above the allowable costs. Then the cost-reduction efforts of "Achieve the Target" come into play. That is the next step, step 3, in the overall Target Costing process, and will be discussed in Chapter 5.

Finally, remember that if you are doing Target Costing that covers a number of years in the future, you must calculate the allowable costs and set subsystem cost targets for each of those years—based on the prices you have determined will prevail in each of those years. This will become clear in the next Module of the Exercise.

**ROBUSTNESS**

The reader may wonder how robust a value analysis approach is. After all, it depends on two or three sets of numbers that have some degree of uncertainty in them. First, the overall target price has some uncertainty since it is a triangulation of experience curves, direct customer input, market surveys, etc. Nevertheless, we have found that—in practice—the projected future prices are quite dependable (say, within ±5%) if they have been determined from a combination of two or three quantitative approaches.

Second, value analysis uses the relative importance percentages of the product attributes or characteristics. These come from customer interviews, Conjoint Analysis, market surveys, etc. Again, if you derive them from quantitative approaches that have direct customer input, these will be dependable and repeatable.

## Set the Target (Subsystem Level) 69

Third, the method depends on the percentage extent to which the subsystems contribute to each product attribute or characteristic. The function–subsystem matrix comes from subjective estimates by system architects and engineers. Generally, this diverse group of individuals will converge quickly on a set of values for the matrix and be able to justify it afterwards.

Finally, the acid test of robustness is to do a sensitivity analysis by changing some of the numbers to see what effect the changes have on the subsystem targets. Naturally, all the allowable-cost results will scale directly with the overall target cost. But the results are fairly insensitive to changes in the percentages of either the relative importance or the function–subsystem matrix. That is because most actual, real-world cases have so many cross-terms that a small perturbation is negligible among many terms in the sumproduct used to calculate the sub-system allowable costs. This holds true as long as the relative importance estimates and the function–subsystem matrix are realistic.

## SUMMARY

In this chapter we have described how to arrive at cost targets for the subsystems. In summary:

- Cross-functional teams are a vital element to have for successful Target Costing.
- A proposed product should be subdivided into its functional subsystems to distribute the target cost, before attempting to find paths to the targets.
- Cost targets are assigned to each subsystem from the relative importance of each feature to the customer.
- These "subsystems" or cost elements can include more than the physical parts of the product, such as manufacturing, testing, procurement, installation, etc.
- Value engineering is a useful way to map features (perceived by the customer) to functions (perceived by the engineer) and to identify candidates for cost reduction.
- Subsystem targets can be validated when there is data to create experience curves at the subsystem level.
- The final cost of any subsystem may exceed its cost target, but the total target cost must be kept constant.
- Experience has shown that target costs are achieved in the vast majority of products, when appropriate resources are applied to the effort.

---

You should now do the next phase of the Exercise.
Please turn to Module D, page 179.

**REFERENCES**

1. Shillito, Larry, et al., *Value: Its Measurement, Design, and Management*, New York, John Wiley & Sons, 1992.
2. Horvath, Peter, Universität Stuttgart, "Target Cost Management in German Companies—a Case Study of Siemens," First Annual International Congress on Target Costing, Conference Proceedings, 1997.

# 5
## Achieve the Target

**WHERE ARE WE IN THE PROCESS?**

We are at the third step in the process, "Achieve the Target," as shown in Fig. 5.1. In the last chapters, we discussed setting the target price and the target cost at the whole-product level, and then set cost targets for the subsystems that comprise the product. In this chapter we will describe ways to find paths to achieve these targets.

At this stage in the Target Costing process, we have done a significant amount of work to understand the customers and the market. We have developed a set of subsystem cost targets that are tied directly to the customers' willingness to pay. This alone is a major accomplishment. Setting a rational target, even if aggressive, usually achieves greater results than "best effort" activities without such a target.

Finding paths to the target depends on clearly defining the problem and then focusing the skills of subject-matter experts on the challenge. Achieving the target depends on unleashing the power of the organizations involved and relies on good project management.

Having targets for the subsystems initiates an assessment of the product design and features. In our experience, initial cost improvements have been achieved by simply examining the product architecture before the in-depth analysis of finding paths is started. Sometimes simple modifications will result in lower cost without compromising the basic features and performance of the product. Some generic cost reduction methods at the architecture level include:

- Providing features in software rather than hardware.
- Using standard subsystems and components in place of custom-designed ones.

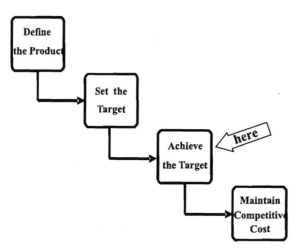

**Figure 5.1** We are at the third step in the process.

- Using subsystems with less capability that still fully meet the customers' set of requirements.
- Changing marginal Basic features to optional Step-up features (but be careful here).

However, these actions may not provide the complete cost savings necessary to meet the targets. In those cases, you will apply the processes described in this chapter. These processes are easy to describe but require conscientious effort to be successful.

### SUBDIVIDING THE SUBSYSTEM COST TARGETS

The process of subdividing the subsystem costs further is illustrated in Fig. 5.2. This is an essential part of the Target Costing process and will generally provide the best strategy for reducing the product's cost to meet the target set in step 2, "Setting the Targets." Simply stated, this activity examines all the parts and functions of each subsystem. Examination of the cost and function of each part will uncover additional opportunities for cost improvement.

Each subsystem usually comprises a number of specific parts or components. We can subdivide the subsystem cost target and allocate it among the several components (illustrated in Modules C and D). Then we work to find ways to manufacture or otherwise obtain each of the components at its individual cost target. This could be done by pro-rataing the difference between the actual and the target cost for the subsystem. But it is better to use methods such as value analysis to apportion the subsystem cost. It is also beneficial to understand the technology road maps for the components—some component types are on much steeper

# Achieve the Target

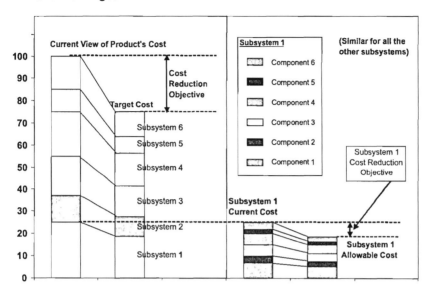

**Figure 5.2** Establishing cost targets for components.

experience curves than others. (Digital signal processor ICs are coming down faster than resistors).

Naturally we recognize that this example might be simpler than a case you might find in your own business. It is often necessary to subdivide many times before getting down to the component or piece-part level. For example, if you were in the shipbuilding industry, the engine is a typical subsystem that you would find in a steamship. But the engine subsystem can be subdivided into several sub-subsystems (e.g., cooling system, combustion chambers, lubrication system, drive train, controls, etc.). Each of those might be further subdivided (e.g., drive train = crankshaft, gear box, drive shaft, etc.). You might go through five or six subdivisions before you get down to the piece-part level.

The essential point is to continue to subdivide the subsystems down to the appropriate component level and to obtain costs for each of the components. Simple Pareto analysis will identify those components that contribute the most cost. The function of each major component will identify opportunities for cost reduction. In some cases a major component has additional capabilities, provided at higher cost, but they are beyond the requirements of the product's functional needs. This type of excessive performance capability leads to higher cost without benefit to the customers. Other examples include functional redundancy, alternative sources, and commercially available components rather that custom-designed components. Product requirements and certifications may also add unnecessary cost. For example, European telephone service providers do not require Underwriters' Laboratories certification (however,

European Standards Testing Institute compliance *is* needed). Having both is not merely a matter of designing and manufacturing to meet these standards, but also includes the expense of actual certification provided by the respective organizations. Having both simultaneously may add unnecessary expense to the product.

## FUNCTIONAL TEAMS

In Chapter 2, we recommended that you establish a cross-functional steering team to start your Target Costing project, to "Define the Product," and to "Set the Target" at the whole-product level. This steering team is made up of key managers from all functional areas. In Chapter 4, we recommended that you establish small functional teams (typically, 3–6 individuals) that focus on each of the subsystems. At this point, these teams can begin to start finding ways to achieve the targets through brainstorming, redesign, supplier interactions, etc. The large group of all functional teams forms the complete cross-functional team. At some points in the process you may further supplement this team by involving subject-matter experts from other parts of your company or from suppliers outside your company.

Each of these teams will have a target cost for their respective subsystem, and they will work to achieve it. The teams will all use the same processes and approaches to generate ideas that reduce the cost. We emphasize that these teams must understand that they are *not* in competition with each other. In fact they must collaborate intensively. A team does not "win" if it finds ways to achieve its target, especially if the overall system cost does not converge on the overall target cost. It is better to take the view that *all* teams win, but *only* if paths are found that achieve the overall target cost. Tradeoffs among the teams are often required. For example, one may find that changing something will increase the cost of one subsystem by $350, but it will reduce the cost of another subsystem by $1250.

It is the steering team's responsibility to manage and coordinate the process of reducing the overall cost, and to orchestrate the tradeoffs among the various teams. Good communications are essential. It is easy for the subteams to diverge from the overall target or reduce the cost of one subsystem at the expense of the other teams. Sharing a clear definition of the interfaces between subsystems during the brainstorming meeting aids close coordination between subsystem teams afterwards. The Steering Team must make tradeoff decisions between subsystems, and the contribution of both subteams in meeting the target cost must be recognized. In one example from our experience, the product's installation cost was reduced by 12% with an increase of only 1% in the manufacturing cost. Overall cost reduction of 5% was achieved. The decision was not easy since one of the manufacturing subteams had to increase its cost when

# Achieve the Target

it was already under pressure for being over its cost target, but the result was best for the larger team and for the customer.

## FINDING PATHS

There are many ways to achieve cost reductions. As suggested above, the first step is to find high-level ways to reduce cost at the architecture level. The next step is to carefully examine the subsystems. The subsystems or other parts and components may be redesigned. Through negotiation, suppliers may be encouraged to offer lower prices. They may suggest replacing items with lower-cost alternatives, and so on. Items can be minimized, rearranged, combined, adapted, substituted, etc. In considering an element of the system, the following short list of questions adapted from Ansari, [1] will help. The examples in parentheses are taken from our experience when we applied Target Costing to outdoor electronic equipment cabinets:

- Is it doing more than the customer requires? (e.g., battery heaters not needed in hot climates)
- If you were the customer, would you omit this? (e.g., decorative door ribs)
- If this were not done, would anything of consequence happen? (e.g., two-tone paint not required)
- Can we interchange steps? (e.g., pre-assemble first, then paint)
- Can we combine functions? (e.g., battery venting and outside air duct)
- Are we trying for too much reliability? (e.g., battery cable wire gauge too large)
- Is there a simpler way of meeting this or most requirements? (e.g., 90° bends)
- Can another firm's standard part or process be used? (e.g., AC power panel)
- Is this the only way to do this? Is this the best way? (e.g., use industry-standard foot pads)
- Do others produce it at less cost or in less time? (e.g., door latches)
- Is there a similar product that meets most of the requirements? (e.g., a supplier's standard cabinet)

## BRAINSTORMING

We have found that brainstorming with all the teams together, and within the individual teams, can generate a remarkable number of cost-reduction ideas. The brainstorming process is enhanced significantly when all known data is available to the participants. Sharing of the data with all participants reduces the amount of time justifying the conclusions of steps 1 and 2 of the process. We start with a review of market issues, the market-feature table, targets and their derivation, and technology road maps. This foundation motivates the team and stimulates appropriate brainstorming by showing team members where their expertise needs to be applied. This approach leads to more focused brainstorming and more useful ideas than jumping in directly to brainstorming.

In our experience with complex telecommunications equipment, we typically obtain 100–150 ideas, and sometimes as many as 250 ideas, after a few hours of structured brainstorming. Earlier we made the point that the teams should comprise subject-matter experts (SMEs) from outside the product group and even from other divisions of the company. These SMEs turn out to be especially fertile sources of new ideas, as they are not constrained by the conventional thinking and assumptions that surround every product group.

All ideas should be accepted when they are first suggested. But once there are a lot of ideas for cost reduction, they need to be assessed. This involves eliminating the clearly impractical and infeasible ones, merging those (from different sources) that are virtually identical, and resolving conflicts among ideas that are mutually exclusive. The ideas that remain then need to be evaluated in terms of (1) the potential savings that they could generate, and (2) the "investment" in people, time, and money needed to put the idea into effect. It may take anywhere from one week to several months to evaluate a full set of ideas and make a final set of recommendations for their implementation.

Most readers have participated in brainstorming sessions. We have found that a facilitator effectively engages all participants appropriately and guides the team so that priorities are met. An experienced facilitator to manage the brainstorming session is one of the most productive ways to get a start on the "Find Paths" in Target Costing.

## BUBBLE DIAGRAMS AND PARETO ANALYSIS

To stimulate effective brainstorming, we use "bubble diagrams" and Pareto charts of the major components and subsystems to reinforce how components contribute to each subsystem. Bubble diagrams are pictures of each component in each subsystem and the relationship of each one's cost to the complete product cost. It is merely a graphical representation of a spreadsheet, but when developed with and understood by a cross-functional team, it is a powerful tool. An example of a bubble chart for a personal computer (PC) is shown in Fig. 5.3. From this chart the relative costs of the subsystems and components are clearly seen. Potential cost reduction, consolidation, enhancement, and elimination ideas will be stimulated by taking each bubble one at a time and examinig its interfaces, requirements, and function, guided by this simple graphic.

Bubble diagrams are magical for getting teams to the component level and stimulating answers to the question, "Why?" With the large cross-functional team together, methodically review each subsystem's

## Achieve the Target

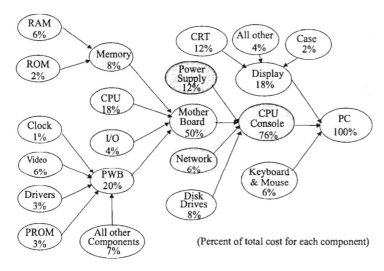

**Figure 5.3** Bubble diagram of typical costs for a simple personal computer.

bubble chart. For each subsystem, ask the team what is involved and draw a bubble. At this lower level, probe deeper with the subject-matter experts for that bubble. Ask them to explain the function and justify the requirements. Then ask the subject-matter experts to quantify the impact and justify it. In the process, someone is bound to offer an alternative. The facilitator should suggest unusual and strange alternatives and ask the subject-matter experts to correct you. This will stimulate discussions and shift the team's thinking far from what was done in the past. Sometimes it helps to stimulate idea generation by challenging the subject-matter experts to consider a very simple approach (e.g., use a paperclip for an antenna) and explain what must be added to meet all the requirements. Often, the simple approach is inadequate, but the original complex approach is overdone and can be simplified.

A different but also useful approach to understanding "Why?" can be found in Chen and Chung's description of cause–effect Analysis [2]. This method can be especially good for determining "Why things cost?," and is very effective when used after a bubble diagram has been constructed.

Another useful approach in finding paths to the target is through Pareto analysis, commonly referred to as the "eighty-twenty rule." Pareto analysis quickly focuses on the major costs of the systems and eliminates the guesswork associated with finding where the biggest "bang for the buck" will be. However, a caution is needed here: cost reductions that impact the decision features as defined in the market-feature table must be carefully examined before acceptance. The value of market-driven design is the totality of the design in meeting both the market requirements for price and performance and the organization's need

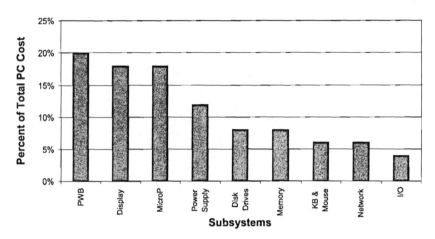

**Figure 5.4** Pareto chart of PC costs.

for profitability. Compromising any of these design constraints will compromise the viability of the product.

Fig. 5.4 presents the results of the Pareto analysis of the cost of the personal computer presented in the bubble diagram of Fig. 5.3. As is usually found in Pareto charts, most of the costs are from a minority of the subsystems. Here the major costs are in the PWB, the display, and the microprocessor. The total cost associated with all the other six systems is less than the cost of the three high-cost systems. The cost of the microprocessor is a single unit and is responsible for 18% of the total PC cost in this example.

The cost-reduction efforts should focus on the high-cost systems. However, in the case of the microprocessor, any changes in this component may materially alter the performance of the PC. However, the technical expertise of the cross-functional team members may identify approaches to reduce the cost of this expensive component while maintaining the features specified in the market-feature table.

Recall from the bubble diagram of Fig. 5.3 that the PWB is composed of several additional components, and cost reductions may be found from analysis of each of these components relative to the features necessary in the market-feature table. The bubble diagram and the Pareto charts present the cross-functional teams with necessary cost data to suggest tradeoffs that can be made among components, to reduce cost and to maintain or increase performance.

## DFX AND TARGET COSTING

DFX (Design for "X") is a design process or discipline that ensures that the requirements of a specific product life-cycle stage ("X") or several stages are addressed and satisfied. Some major product life-cycle stages

# Achieve the Target

are Procurement, Manufacture, Test, Assembly, Configuration and Installation, Maintenance and Repair, and Recycle. Thus DFM is Design for Manufacture, DFT is Design for Testing. "X" can also refer to key product issues such as Safety (DFS), Environmental Compliance (DFE) and the like. Each of the DFX processes has a set of guidelines that assist a design team in creating a product that will deliver the performance specifications and will optimize the product life-cycle processes.

In the Target Costing process, DFX is one of the tools that can be used to help achieve the product's target cost, and this is especially true if "cost" embraces full-stream costs. DFX is applied in the phase "Achieve the Target" (see Fig. 5.10). DFX is very useful for identifying cost-reduction activities beyond the efforts of reducing COGS—the cost of components and materials. A simple example of DFM is reducing the cost of manufacturing by designing the product to be assembled with a minimum of work-piece movement, or a reduction of fasteners. DFI (Installation) reduces costs through simplifying the installation process, a major cost in many industrial products. DFR (Recycling) can reduce costs for those products that must be recycled, or where the cost of disposal of the product and/or its constituents is an issue.

A detailed treatment of DFX is outside the scope of this work, but the usefulness of DFX is very important in reducing direct costs, shortening the manufacturing interval, simplifying the installation and test process, and so on. See the references [3–6] for additional background and processes in DFX.

## SUPPLIER TARGET COSTING

As you work deeper into the bubble diagram, a component or subsystem may turn out to be an item that is entirely purchased from suppliers. This provides an opportunity to work with those suppliers to help them reduce their costs, with the expectation that some or most of the savings will be passed along to you as price reductions. Suppliers can help identify cost issues in your component or subsystem specifications that are adding costs without significant benefit. They can also suggest alternative approaches that provide an adequate capability at a lower cost. Furthermore, suppliers can apply Target Costing within their own firms to the products they sell to your company. In this case, your firm is their customer. You show the value analysis so your suppliers understand the rationale for the cost (their price) targets that you establish for the item(s) they provide.

In the overall Target Costing process, this is where your firm engages its suppliers. It can also be an opportunity to learn from suppliers and validate the targets and design choices you have made. These brainstorming sessions with suppliers (a typical agenda is shown in Fig. 5.5) are often preceded by a tour of the supplier's facility and capabilities. Suppliers appreciate the opportunity to show their expertise

> **Welcome & Introduction**
> >   Target Costing Process Overview
> >   Team Introductions
>
> **Marketplace Issues & Challenge**
> >   Customer Requirements, Relation to Other Products
> >   Target Cost Derivation, Business Objectives
>
> **Design Requirements**
> >   Requirements
> >   Subsystem Breakout & Interfaces
> >   Subsystem Target Costs, Volume
>
> **Supplier's Latest Proposal**   (as whole team or separately, by sub-system)
> >   Product Line Evolution
> >   Breakout to Component Level  (bubble diagram for key components)
> >   Prepared Proposals
>
> **Brainstorming**
> >   Idea Generation by Subsystem or Component
>
> **Prioritization of ideas**
> >   Classification
> >   Assign Action Items

**Figure 5.5**   Typical supplier brainstorming meeting.

and to better understand the requirements in order to offer better solutions. Depending on the level of engagement and depth of the relationship with the supplier, a discussion of technology or product road maps before brainstorming may allow breakthroughs that help both parties.

If multiple suppliers are involved, the environment and social dynamics are different. First there is usually not an opportunity to tour facilities, but a presentation can cover the highlights. In some cases the suppliers may not customarily work together, but are brought together by Target Costing to collaborate on approaches where they all contribute to the solution to meet the target cost. This works best when a major subcontractor participates with its key suppliers, your firm's second-tier suppliers. This provides the second-tier suppliers a golden opportunity to understand their customer's needs, creating a win-win situation for all involved.

In other cases, the suppliers may be direct competitors because your firm has multiple suppliers for the same subsystem. Surprisingly, many of the same steps can be followed if planned appropriately. Here the review of market issues, target derivation, and technical requirements is presented simultaneously to all suppliers. Then subteams are formed where competitors are put on separate teams. In this way, participants can discuss approaches and brainstorm with other subject-matter experts

## Achieve the Target

with whom they are comfortable. In truly competitive situations, where engagement is early in the product-realization process, this accelerates the market forces. This approach can be used to determine which subteam best meets the requirements and influence supplier selection decisions.

### CLASSIFICATION AND FOLLOW-UP ON IDEAS

Once ideas are generated, the next step is to evaluate them. The first level of review is at the end of the brainstorming meeting. When ideas are first presented it is important not to criticize them. The second time through, they are discussed and evaluated. When the team is together, you have a golden opportunity to use the collective wisdom of the cross-functional team gathered for the meeting. For each idea, read it out loud and ask someone to explain what it means. Usually the originator will explain. Then once everyone understands the idea, ask the team whether it has a "high benefit" or a "low benefit." Sometimes this is obvious immediately, but many times it is not. First a team must agree on the difference between high and low. Sometimes they agree on a savings value or percentage, sometimes it is an amount of time saved, or something else.

Once definitions for "high" and "low" are settled, the team must reach consensus on the benefit classification. With a cross-functional team, the different perspectives may have a different view. What is "high" for one stakeholder may be "low" for another, especially if suppliers are participating. When there is a disagreement, the opposite sides present their rationales, and that leads to a discussion of the advantages and disadvantages of the idea. After some discussion, a facilitator guides the team to agreement. Once the team agrees on benefit, it is time to turn to implementation effort. Here the choices are "easy to implement" and "difficult to implement." Again, a diverse team may not agree on the difference and will bring up pros and cons before reaching consensus. This is especially insightful if a customer and supplier participate on the cross-functional team. Fig. 5.6 summarizes the classification matrix. Systematically discussing each idea in this manner gives them each a convenient classification name based on value.

Usually at this point there is enough support to pursue some of the category 1 ideas. Other ideas need further evaluation or validation that will take time and resources not available at the meeting. Therefore, time should be taken at the end of the meeting to capture the key points of the classification discussion. Fig. 5.7 is a good form for capturing the key points from the discussion for sharing with others. Following the meeting, participants follow up on missing information, test hypotheses, prepare quotes, validate assumptions, etc. Soon afterwards the team meets to review the status of action items, reconsider the ideas, and, when there is enough information, decide whether to pursue or reject them.

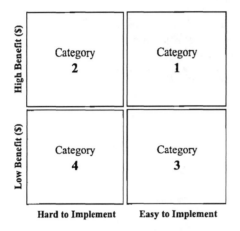

**Figure 5.6** Idea classification matrix.

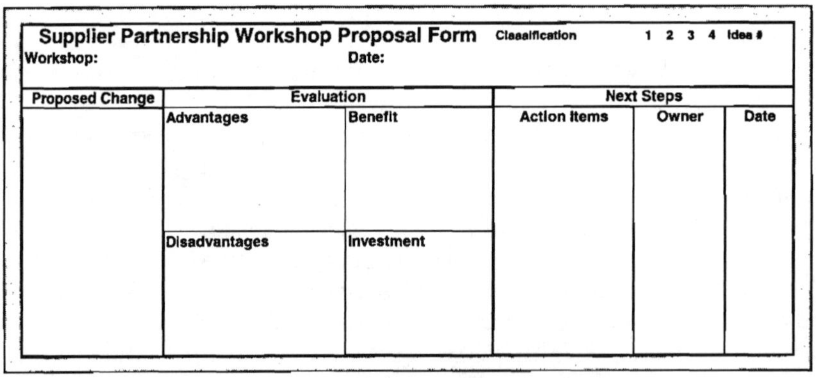

**Figure 5.7** Proposal form to capture brainstormed ideas.

## ACHIEVING THE TARGET UP THROUGH PRODUCT ROLLOUT

Once ideas are generated, it will take some time to evaluate them. In some cases, more investigation is required to validate assumptions made during the brainstorming meeting. The steering team leads the team through evaluation of the ideas after the brainstorming session and coordinates tradeoffs between the teams as adjustments are made. During this time, market conditions and competitors' actions are likely to change and can affect the decision on particular proposals. It is important to review the data and assumptions used in formulating paths to the targets in the context of the latest views of the marketplace. For example, a change in the ecomomy may change the volume used to estimate the payback of a specific proposal.

### Achieve the Target

Once feasible paths to achieve the target are found, you must still successfully implement the recommendations in such a way that actually achieves the target cost when the product is introduced into the marketplace. Failure to implement the cost-reduction ideas is one way to miss the target cost. Disciplined project management and cost tracking will guide the team to successful implementation of the proposals.

"Feature creep" is another phenomenon that can cause a team to miss the target cost. This occurs when it is determined that additional features should be added to the product during late development stages. Often these additional features are inserted by Sales or Product Management, based on perceived pressures from customers. They may be responding to new products that alter customers expectations, and thus the basic decision factors. What is frequently missing is the idea that if these features are truly valuable to the customers, they should be willing to pay for them. A determined effort must be made to establish the value (hence the price) of these new features. If all customers want them, they can be added to the "Basic" product, the target price can be adjusted upwards, and so can the target cost. If only a subset of customers want the new features, then they should be considered "Step-Up" or "Premium" and made available at an optional price.

Technology may also change the alternatives available. For example, the introduction of broadband Internet provided alternatives to T1 phone lines for high-speed data tranmission. In another example, mobile phone service prices have become competitive with fixed-wire phones. This caution should not impede the decision process, but it ensures due diligence in the process.

### THE END GAME

The steering team must periodically evaluate progress toward the overall target cost. The best situation is when the sum of the efforts of all the subsystem teams is such that the overall target cost can be achieved. If one or more of the teams is having particular difficulties, increased attention to that part of the system may be necessary. If the overall target cost is still not being approached, then strategic decisions must be made, as summarized in Fig. 5.8. The product might still be introduced for strategic reasons, such as rounding out a product portfolio, but it will be at a cost disadvantage. If the product is not strategic, the product development should be terminated.

Recognize that not every company can develop every product. There are limits on resources, expertise, access to customers, market share, etc., that sometimes prevent a team from meeting the target cost. In these cases, the financial health of the company is better served by abandoning the project. A poorly designed product, or a product that does not meet the market's requirements, will fail. Just as it is best to start early to

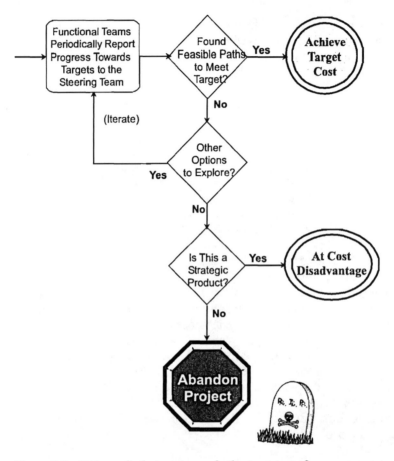

**Figure 5.8** Did you find ways to reach the target cost?

achieve the target, it is also better to abandon a product development at this early stage rather than after significant costs are incurred. The lesson is to cancel a failed project as soon as you know it will not succeed.

## A WORD ON PROCESS

Fig. 5.9 shows the details of the steps associated with the "find paths" phase of "Achieve the Target." It indicates that, after a brainstorming meeting, each of the subsystem teams works to achieve its subsystem cost target. As we said before, the teams also work with each other to check overall progress, to make tradeoffs, and to ensure that the overall target cost can be reached. At a readout meeting the results are presented to executives or senior managers, along with indications of required incremental investment and implementation recommendations.

# Achieve the Target

## Inputs:

(Work mainly with Design, Architects, Systems Engineers, Manufacturing, Installation, Purchasing, Supply-Chain, Finance, Product Managers, and Subject-Matter Experts outside the business group.)

- Confirm Exec. "buy-in", commitment.
- Review business-group strategy.
- Do Tax review.
- Generate Baseline View.
- Identify Teams & Members.
- Do Peer Review.
- Hold Brainstorming Meeting.
- Validate, evaluate and estimate ideas.
- Have Readout with business-group stakeholders; show paths to the targets, investments required, and recommendations.

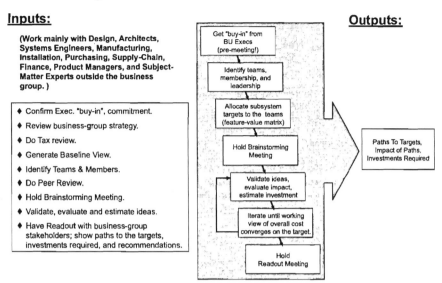

## Outputs:

Paths To Targets, Impact of Paths, Investments Required

**Figure 5.9** The steps associated with "Achieve the Target" (finding paths phase).

Fig. 5.10 shows the details of the steps associated with the phase of "Achieve the Target" after the brainstorming and readout meetings and before the rollout of the product.

## SUMMARY

In this chapter we have described a number of aspects about finding paths to achieve the targets. Some of the main points are:

- Use cross-functional teams to further subdivide the product into subsystems and components.
- Continue using value engineering and brainstorming about potential subsystems and components as redesign candidates and to look for adaptations, combinations, rearrangements, and substitutions.
- Analyze the most promising cost reduction ideas; use Pareto analysis. Selected ideas must be valid (technically feasible and acceptable to the customer).
- Re-examine the feature set if costs cannot be met. Some features may have such a low value index that they may be eliminated and greatly reduce the cost.
- Develop selected ideas into concrete cost-reduction proposals.
- Identify and verify the workability of risky issues.
- Identify the set of feasible paths to achieve the targets, with their cost impact and implementation resources required.

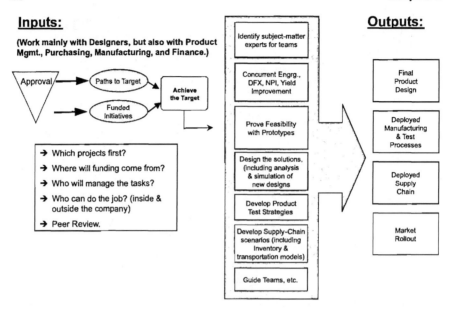

**Figure 5.10** The steps associated with "Achieve the Target" (up through product rollout).

- Use disciplined project management and tracking to assure that the product remains "on target" through its introduction into the marketplace.

> You should now do the next phase of the Exercise.
> Please turn to Module E, page 191.

## REFERENCES

1. Ansari, S. L., Bell, J. E., et. al., *Target Costing: The Next Frontier in Strategic Cost Management*, Consortium for Advanced Manufacturing International (CAM-I), Chicago: Irwin, 1997, pp 132–133 (most recently listed McGraw-Hill as publisher).
2. Chen, Richard C. and Chung, Chen H., "Cause-Effect Analysis for Target Costing," *Management Accounting Quarterly*, Winter 2002, pp. 1–9.
3. Boothroyd and Dewhurst, "Product Design for Manufacture and Assembly," Manufacturing Engineering, April 1988.
4. Boothroyd and Dewhurst website, http:/www.dfma.com. Includes information on design for X.

5. Maunder and Beenker, "Boundary Scan: A Framework for Structured Design fot Test," Proceedings IEEE International Test Conference, 1987.
6. Rehg and Kraebber, "Computer Integrated Manufacturing," 2nd ed., Prentice-Hall, 2001.

# 6
# Maintain Competitive Costs

**WHERE ARE WE IN THE PROCESS?**

We are at the fourth and final step in the process, "Maintain Competitive Costs," as shown in Fig. 6.1. In the previous chapters, we discussed setting the target price and cost, and achieving the target cost up through the rollout or introduction of the product. In this chapter we will describe steps that you should take to maintain competitive costs through the life cycle of the product.

You learned in Chapter 3 that prices and costs for equivalent functionality inexorably decline (in constant dollars). This is the result of the cumulative experience of the industry with competitive pressures, reducing costs, and implementing new technologies.

The approach provided here still uses the cross-functional teams, but allows members to retain their affiliation with their home organization and fellow subject-matter experts. For example, accountants stay in Finance, while engineers work in their respective development groups. This may be extended if you are involving suppliers as virtual members of your company because it allows them to stay in their respective companies!

**A FRAMEWORK FOR THE TEAMS**

Paths to the targets are generated at all levels of a company. Many of the same groups generating the ideas also develop and deploy them. However, before they can start, they usually need agreement from other stakeholders. This requires an organizational framework where each discipline has clearly defined roles and responsibilities to coordinate the teams' efforts to "Achieve the Targets."

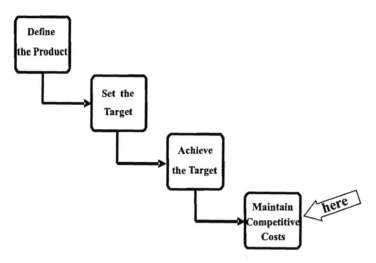

**Figure 6.1** We are at the final step in the process.

Achieving cost targets is driven from the top of the company. An overall objective for the Chief Operating Officer like "achieve target costs for the products that contribute 80% of the revenue" works wonders in getting alignment of the other teams. This objective unifies the supporting cross-functional team for a product family that includes representatives from Product Management, Sales, Competitive Analysis, Development, Purchasing, Manufacturing, Installation, and Finance. The office of the COO ties these together.

To tie them together, organize the company's product portfolio by product family. At regular intervals, have subject-matter experts from the various functions meet to make plans and review the results. These reviews include a discussion of ideas generated to achieve the target and agreement on priorities. We recommend a set of standard templates describing the relevant issues for each product family to unify these groups in their common goal to "Achieve Target Costs."

For each product family, select versions of the product called *product configurations* that represent what is sold. Typically these are a popular version of the product representing the "Basic" model used in setting the targets. One could also represent a typical or an average product.

The example in Fig. 6.2 is the centerpiece of the reviews. It is the first of three templates that we have found to be useful in maintaining the target. It ties together the target cost, current cost, product cost plan, and key product improvements. There should be one for each major configuration in the product family. The more you choose, the better you represent the total product family, but the more complex your model and potentially greater effort is required. Usually there is much re-use of parts and commonality between configurations, so two to five are sufficient for even the most complicated product families and markets.

## Maintain Competitive Costs

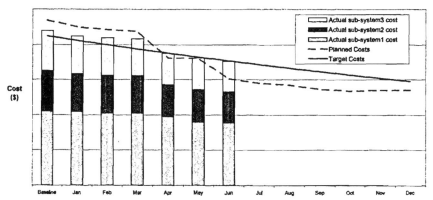

**Figure 6.2** Template 1 example, used for tracking costs during the "Achieve the Target" phase.

Each organization has a role in achieving the company's overall objective of "Achieving Target Costs" and contributes to creating Template 1 for each configuration. The roles and connection to Template 1 are as follows:

- *Product Management*—defines the configurations that meet requirements for market "sweet spots" and sets priorities for development and release of product features.
- *Sales*—provides market forecast for product configurations. The price and quantity determine the size of this business for the firm and thus the relative importance of a product configuration in the company's portfolio.
- *Competitive Analysis*—provides cost targets required to meet financial objectives for the identified configurations.
- *Development*—provides product architecture alternatives and develops initiatives to meet product plans.
- *Purchasing*—engages suppliers who provide materials, manufacturing methods, and logistics approaches.
- *Manufacturing*—provides input to design, produces the product, manages the inventory, and keeps track of the actual manufacturing costs.
- *Installation*—provides input to design, installs the product, and keeps track of the actual installation costs.
- *Finance*—provides financial results closing the loop on initiative plans.

The other recommended templates support this one and provide the details needed to make key decisions. Template 2 in Fig. 6.3 is a list of approved ideas, their impact, timelines, and resource requirements. Part of the meeting should be spent reviewing new ideas to see which ones should be funded and implemented. By reviewing the ideas in the context of all the configurations, the team has the information needed to prioritize the initiatives based on their impact across multiple products and resource availability. Once approved, staffed, and funded, time should be spent on status reports covering development and deployment issues. When deployed, Finance can provide an accounting of whether financial objectives were met.

Template 3, in Fig. 6.4, is an additional way of looking at the impact of deployed initiatives. It shows the savings plan for an existing product family by initiative for a fiscal year. This comes into play when you are

| Team | Idea / Initiative # | Description | Impact Date | Savings/unit | Total FY Plan | Investment Required | Actual Results |
|---|---|---|---|---|---|---|---|
| Design | 1 | | | | | | |
| | 2 | | | | | | |
| | 3 | | | | | | |
| | 4 | | | | | | |
| | 5 | | | | | | |
| Purchasing | | | | | | | |
| Manufacturing | | | | | | | |
| Installation | | | | | | | |

Results for each initiative validated by Finance, closing the target-costing loop.

**Figure 6.3** Template 2 example, used for tracking progress of brainstorming ideas.

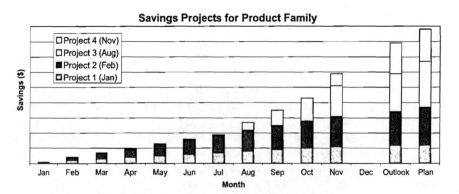

**Figure 6.4** Configuration level savings summary by month, as viewed in November.

# Maintain Competitive Costs

tracking the progress of initiatives associated with a complete product family. Many of the initiatives can affect the costs of more than one configuration in a product family, and they may have different starting dates, so their combined impact is shown in this way.

## MAINTAINING COSTS

Maintaining competitive costs implies that one has a cost plan to follow. The cost plan is developed from the sum of the sales of the products in the different regions. These individual product cost plans come from the price trend and the required profit margins—in the ideal situation, the target cost for each! Note that profit margins vary from product to product depending on customer, region, and stage in the life cycle of the product and that the margins usually decline over time with competition.

Once you have a cost plan, you will want to monitor your progress toward achieving it. The first three templates are useful for tracking progress toward achieving the target cost for a single configuration. However, in most cases Templates 1 through 3 are *insufficient* for maintaining costs because the total product line consists of much more than the two to five configurations selected to represent it. There are usually spare parts, options, and other low-volume configurations that are not analyzed and studied in the same detail as the "Basic" models. A fourth template provided by Finance is needed. Template 4 (example shown in Fig. 6.5) is a total cost plan for each product family. Total cost is the sum of volume times unit costs for all products. It includes the important configurations that are analyzed in detail *plus* other product sales, including lesser configurations, spare parts, units for repairs, and ancillary materials that support the primary products. It insures that the configurations selected represent what is actually happening to the product

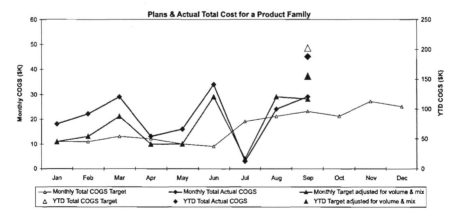

**Figure 6.5** Template 4 example, used for comparing a product family's actual total costs to total cost plan.

family in the market. If there are significant differences in the sum of the key configurations and the totals, then either more configurations are needed or some key ones are missing from the detailed analysis. This template also shows whether the initiatives have impact beyond the key configurations for other parts of the portfolio.

In this example, the total COGS targets were worked out for the entire product family based on assumptions for sales volume and mix by region. So this represents the *total* COGS for the family, not the COGS of one product within it. The light triangles show the plan by month for the year. The diamonds show what actually happened. If you merely compared the actual to the plan, you would miss two important factors, changes in volume and changes in mix. Therefore the dark triangle shows the original plan normalized by a ratio of actual volume to forecasted volume. You could also normalize for changes in mix, but it was not necessary in this case since actual mix tracked well with planned mix. In this way, we can see if we are staying on track, even if the sales volumes and mix are different from the forecasts.

Once a tracking mechanism is in place, you have to exercise the discipline to use it. You also must be prepared to take action when the actual incurred costs drift above the cost track that you have established. Remedial action, as always, involves identifying root causes, proposing remedies, and implementing improvements.

## PORTFOLIO LEVEL REVIEWS

The templates provided so far summarize plans and results at the product family level. They may also be used in reviews with the Chief Operating Officer seeking to achieve target costs company-wide. Template 5 (Fig. 6.6) graphically depicts key summary data from Template 1 for many product

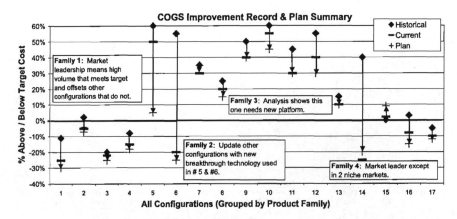

**Figure 6.6** Template 5 summarizes cost improvement plans compared to targets for a portfolio of products.

## Maintain Competitive Costs

families all at once. Each vertical arrow depicts the ratio of actual and planned cost to its target for one configuration as a function of time. Recall that the % above or below is calculated as follows:

% Above or below target = (Your company's cost/Target cost) − 1

The dark horizontal line in the middle of the chart at 0% is the point at which the target cost is met. The historical, current, and future costs are marked along the arrows. A quick glance dramatically shows those product configurations that have met or will meet the target cost and those that need attention. Recall that targets get more challenging over time. If a product's arrow points down, costs are going down faster than the target cost experience curve, so the configuration's cost is improving with respect to the target. If the arrow points up, it usually means there is severe competition and little or no improvement plans for that configuration.

This example from an electronics company shows eighteen configurations across four product families. A team's familiarity with the product configuration or market situation will reveal other points that can be added to the chart. For example, Family 1 has a market leadership position. Recent improvements have brought costs for all four major configurations below the target. Since volume is high, this family offsets some others that are struggling to meet targets. Family 2 has many configurations for similar markets. Configuration 7 has just undergone a major architectural change implementing breakthrough technology. Configuration 8 serves a similar market and is losing ground so might be discontinued. All configurations for Family 3 are far from the target. Plans are needed to close the gap. A recent project brought all major configurations within Family 4 to best-in-class cost levels. The remaining two serve niche markets and thus are not as critical for the company.

To drive overall portfolio cost improvement, we have found the metric % portfolio at target cost effective. To quantify the overall cost performance of a product portfolio, approximate it by summing the contribution of each configuration in the portfolio that meets the target cost with the following equation:

$$\% \text{ Portfolio at target cost} = \sum_{\text{for all configs. at target cost}} (\% \text{ COGS})$$

### QUANTIFYING THE GAP

Percentage gap has its place, but we need to take this further. One must also include the magnitude of the target cost gap. A target cost gap is useful for setting priorities. Typically one chooses a standard time period like one quarter or one year's worth of production. It is impossible to get a precise volume forecast for everything to be sold in a year, so assumptions must be made. Usually revenue and gross margin forecasts are available and they can be combined with the data we already have for the configurations to arrive at a reasonable approximation of the value of closing the target cost gap. Fig. 6.7 shows a spreadsheet tool that can be used to calculate the value of closing the target cost gap. The opportunity of closing the gap is usually compared to the cost of resource requirements to implement initiatives in a cost/benefit analysis.

In the first column, list the names of all the configurations for each product family. In the second and third columns, provide the total COGS plan and portion represented by the configuration, respectively. The remaining columns are for % above or below target cost at various times. Estimate the value of closing each target cost gap as follows:

$$\text{Gap} = C * [P_{above}/(1 + P_{above})]$$

where $P_{above}$ is % above target for that configuration and $C$ is the total COGS represented by that configuration. This approach makes the magnitude as well as the % gap visible for decision-making using available data. Of course you can merely sum the gap for each configuration if you have them.

---

**For Practitioners: The Mathematics Behind the Spreadsheets.** Some may ask how was this formula derived; we provide the derivation here. The ratio of the original COGS plan, $C$, to the COGS plan less the target cost gap, $g$, can be expressed as

$$\frac{C}{C-g}$$

and is proportional to the ratio of the current cost in percent to the target cost (100%), so

$$\frac{C}{C-g} = \frac{1 + P_{above}}{1}$$

So solving for $g$ algebraically, we get

$$C = (C - g)(1 + P_{above})$$
$$= C + CP_{above} - g - gP_{above}$$
$$g(1 + P_{above}) = CP_{above}$$
$$g = C\left(\frac{P_{above}}{(1 + P_{above})}\right)$$

---

In the hypothetical example of Fig. 6.7, there are two product families, each represented by three configurations. The total product

# Maintain Competitive Costs

| | Total Plan | Selection Name | Total Selected | % above target weighted by contribution to total | | | | Total Gap |
|---|---|---|---|---|---|---|---|---|
| | $ 500,000 | config | 90% | 35% | 40% | 40% | 60% | $ 21,522 |

| Product/ Configuration Name | Annual COGS Plan | % Total Product COGS | Select Variable | Portion Selected | % above target | | | | Remaining Gaps (4th quarter) |
|---|---|---|---|---|---|---|---|---|---|
| | | | | | 1st Quarter | 2nd Quarter | 3rd Quarter | 4th Quarter | |
| **Product Family** | | | | | | | | | |
| **Product A** | | | | | | | | | |
| Configuration 1 | $ 125,000 | 25% | config | 25% | 15% | 10% | 5% | 0% | $ - |
| Configuration 2 | $ 100,000 | 20% | config | 20% | -15% | -12% | -10% | -5% | $ - |
| Configuration 3 | $ 75,000 | 15% | config | 15% | 25% | 25% | 25% | 25% | $ 15,000 |
| **Product B** | | | | | | | | | |
| Configuration 1 | $ 75,000 | 15% | config | 15% | -5% | 0% | -10% | -10% | $ - |
| Configuration 2 | $ 50,000 | 10% | config | 10% | 10% | 15% | 15% | 15% | $ 6,522 |
| Configuration 3 | $ 25,000 | 5% | config | 5% | 5% | 0% | -5% | 0% | $ - |
| **All Others** | $ 50,000 | 10% | other | | | | | | |

**Figure 6.7** Spreadsheet for summarizing all configurations and calculating the value of closing the target cost gap.

COGS is $500,000. Only $50,000 is not represented by the configurations, so 90% of total product COGS is tracked with a Template 1 cost analysis road map chart. Overall the % above target is improving throughout the year for most of the configurations. At the beginning of the year only 35% of COGS are at the target cost or better. In the second quarter, Product B's Configuration 3 reaches the target. Since it represents 5% of the total COGS, the % at target cost improves to 40%. In the fourth quarter, Product A's Configuration 1 reaches the target, improving overall portfolio metric to 60% at target cost, weighted by COGS plan. At the end of the year, two configurations are not at the target cost, for a total gap of over $20,000.

This approach for dealing with target cost gaps provides:

- Tracking of key parameters for each configuration representing all major platforms.
- Overall portfolio performance metric using weighted average at target cost based on product sold.
- Visibility of best improvement opportunities for each product family.
- Framework for portfolio level tradeoffs so cost-reduction resources focused where most needed.
- Cascade down to subsystem level and lower to engage subject-matter experts in specific cost-reduction initiatives.

## REVIEW MEETINGS

Once the templates are in place, you must exercise the discipline to use them. You also must be prepared to take action when the actual incurred costs do not meet the plan. Remedial action, as always, involves identifying root causes, proposing remedies, and implementing improvements. The best way to get visibility and support for actions is to conduct regular review of the data. In the review meetings, have product managers present and summarize their results and plans for the product families. Fig. 6.8 shows a typical agenda.

In maintaining costs of the portfolio as a whole, this process:

- Tracks the portfolio's costs and quantifies future opportunities.
- Shows discontinuous improvements arising from major product changes.
- Identifies and helps prioritize best opportunities.
- Helps decide where to make investments and what resources to apply.
- Gets alignment and support across the business group, or the company as a whole.
- Indicates savings achieved when goals are linked to external targets and visible to all.

## Maintain Competitive Costs

**Introduction**
   Executive Summary
      Improvement Plan Summary (Template 5 - figure 6.6)
      Highlights (e.g. % target, savings metrics, product issues)

   Higlight topic of the day
      (e.g. fiscal year business plan,
      new or discontinued products,
      new savings initiatives, inventory issues, etc.)

**Total Portfolio Review**
   Portfolio Overview of CWV vs. Target (figure 6.5)
   Financial Results Overview (totals of product level results)

**Product Level Reviews**
   Volume, Forecast, Inventory situation
   Product Target Costing Summary (Template 1 - figure 6.2)
   Initiative Plans & Results (Templates 2 & 3 - figures 6.3 & 6.4)
   Product Family Results (Template 4 - figure 6.5)
   Issues (e.g. market conditions, resource shortages, delays)

**Figure 6.8**  Typical cost review meeting agenda.

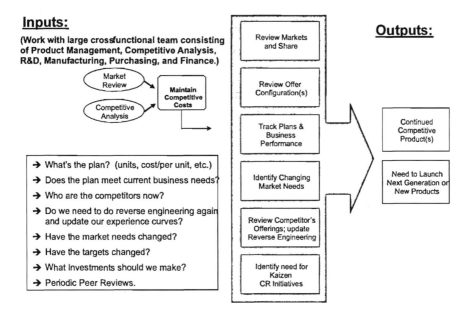

**Figure 6.9**  The steps associated with "Maintain Competitive Costs."

## A WORD ON PROCESS

Fig. 6.9 shows the inputs, questions to ask, and details of the steps within "Maintain Competitive Product Costs."

## SUMMARY

In this chapter we described a framework for maintaining competitive product Costs. Some of the main points are:

- Management support is essential.
- Project plans must have adequate funding for R&D, Purchasing, and Manufacturing initiatives.
- Initiatives should fit into an overall strategy.
- It is important to continue ongoing competitive analysis and reverse engineering.
- It is important to ensure that material costs stay on industry curves.
- Ongoing reporting using standard templates should be used to highlight issues early.

> You should now do the next phase of the Exercise.
> Please turn to Module F, page 211,
> and then Module G, page 219.

# 7

## Putting Target Costing into Practice

**WHAT MORE IS THERE?**

At this point in the book you have an intellectual understanding of the logic and discipline of Target Costing. You realize that it is a market-driven, quantitative, disciplined process for determining what your customers want and are willing to pay for, and for arriving at a price that your customers will like and at costs that allow you to be profitable. For example, in Fig. 7.1 we show portions of the first three main Target Costing steps (Define the Product, Set the Targets, and the "Find Paths" part of Achieve the Target), with the second step expanded. Fig. 7.1 indicates many of the tools and techniques that you can apply along the way.

You also have an intellectual understanding of what steps you need to take as you proceed through a Target Costing project. The Exercise gave you a hands-on feel for going through the steps, and illustrated how one step leads logically to the next. But that was for a contrived case, and you are probably wondering, "Yes, but how do I apply this in the 'real world,' in my real job?" Indeed, the steps, the tools, and the techniques are comparatively easy to describe. Up to now, in the text and in the Exercise, we have focused chiefly on the "mechanical" aspects of Target Costing.

But in the "real world" there are other factors that you will encounter when you try to put Target Costing into practice. Many of these are human and organizational—the so-called "soft" issues. You will be advocating a different way of doing things, and that frequently results in apathy or even resistance. You will have to be aware of the important roles that many individuals play in the organization, and you will have to enroll their support in order to implement Target Costing.

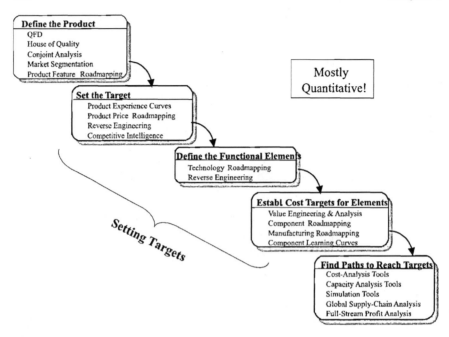

**Figure 7.1** Some principal Target Costing tools and techniques.

In this chapter we provide you with convenient checklists and advice that should make it easier for you to successfully complete your first few Target Costing projects. After that, the results will speak for themselves.

## ARE YOU READY?

One of the first things you should determine is if there is even a basis for applying Target Costing to a proposed product. We have noted in our experience that occasionally there isn't a very sound basis on which to begin a Target Costing project. If a project is proposed, we have found the series of questions in Fig. 7.2 to be useful. (No doubt you could add or substitute similar questions of your own.) Most (but not all) of these questions should be directed at the product managers and marketing people.

We suggest a quick first pass in which you don't try to get the specific answers to the questions, but you determine only, "Yes or no? Do you have the answer to this question right now, or can you get it very quickly?" If there are a lot of "Yes"s to the questions, then the product is probably ready for Target Costing. In that case you next have to start digging and get the specific answers to the questions, because you will need them as you put together the qualitative and

## Putting Target Costing into Practice

**The Need for the Product:**
1. What market need is served by the product?
2. Are there alternative solutions for the market need?
3. What advantages does this product bring to the marketplace?

**Market Strategy:**
1. What are the Company's objectives in this market?
2. How does this product fit into the Company's overall strategy?
3. Are there alternative strategies?
4. Is the planned investment sufficient to cover the risks?

**The Market for the Product:**
1. What are the market drivers for this product?
2. Who are the customers and where are they?
3. How does it affect the customer's Total Cost of Ownership?
4. What are their "wants" and "needs"?
5. How do they differ from customer to customer?
6. How do requirements differ from region to region (i.e. NAR vs. International)?
7. How is the market segmented?
8. Where is market growth?
9. Which markets will be addressed by the Company?

**Price:**
1. What are historical and projected prices (price vs. cumulative units) for this product?
2. What are the customers willing to pay, as a function of features and options?

**Competitors:**
1. Who are the competitors and what are they offering?
2. What Competitive Analyses have been done? Outside studies? Internal? Reverse Engineering?
3. If you were a competitor, how would you compete with this product?

**The Product:**
1. Has a QFD, Conjoint Analysis &/or Price Sensitivity Analysis been done? Results?
2. What are the product features? (Distinguish by "must have" & "desirable to have".)
3. What does the Market-Feature Table show? (What's "basic", "step-up", "premium"?)
4. Is there a technology track for this product?

**Architecture:**
1. What are the subsystems of the product?
2. What are the interfaces between them?
3. Is the architecture well matched to market-feature table?

**Costs:**
1. What cost elements must be considered? (COGS? COGS+? Full-Stream?)
2. What are the required margins as a function of time?

**Miscellaneous:**
1. What other significant items or issues are in the business case?
2. What are the dependencies? (technology? development? other organizations? ...?)

**Figure 7.2** Are you ready for Target Costing—do you have answers to these questions (yes or no)?

quantitative information that you need to use. If, on the other hand, there are a lot of "No"s, then you probably aren't ready to consider launching the product.

## CHAMPIONS

We can't emphasize enough the need to have one or more high-level "champions." One of your first tasks should be to ensure that you enroll one or more champions who:

- believe in the value of Target Costing;
- believe in the need for it;
- give visible public support for it;
- ensure that the necessary people are assigned to work on the Target Costing project;
- will clear roadblocks, if necessary.

## WHOM TO INVOLVE

Doing Target Costing requires time and effort—not just your own. You will require the participation of many people in your organization. Some of them will only have to provide small amounts of input and information, while others may have to spend several weeks doing detailed technical, design, or engineering work. You will also need the participation of your champion(s). Every company, large and small, is different and titles may not translate well from one to another. Small companies may have only one person responsible for a function, or even one person performing multiple functions. In those cases it is easy to identify the key stakeholders. Larger companies may have entire organizations performing some functions, and your challenge will be to identify the key stakeholders related to your project. Examples of the people or functions that you might involve are:

- higher-level champion
- product management
- sales customer representatives
- competitive analysis
- roadmapping
- product architecture
- product systems engineering
- product designers
- manufacturing and test engineering
- component engineers
- installation and service
- purchasing and supply chain
- cost analysis
- subject-matter experts from outside the immediate organization.

## TIME AND EFFORT

Fig. 7.3 indicates the first three principal steps in the Target Costing process, again with "Set" the "Targets" expanded. The figure shows some of the information that you should have at each point. It also suggests that a smoothly run project typically takes three months.

## THE FRONT END

Once you have identified the project, and obtained buy-in by one or more executives or champions, you need to Define the Product and Set the Targets. One or two Target Costing people can typically do this in 3–6 weeks, as long as there is good cooperation, and information exists and is readily available. The work should be done in close collaboration and partnership with the principal product manager(s) responsible for the product. This phase of the work involves a lot of one-on-one interactions with people who know about the product, its markets, its competitors, and its initial architecture. The Target Costing person(s) should touch base frequently with the product manager(s) to validate the information, check assumptions, and agree to the evolving body of Target Costing findings and recommendations. At the end of this phase (the fourth box in Fig. 7.3), you should have cost targets and a sound basis for them. You should also know whether or not the present plans are on a path to achieve the targets.

## BRAINSTORMING MEETING

If there are substantial gaps between the market-based target costs and the plan of record (and there usually are), you need to move to the next steps. The next step—the brainstorming meeting—is always interesting, entertaining, and satisfying. The results are usually more than expected. Prior to the meeting, the stakeholders have to be identified (many of the people suggested by the list on page 104) and invited. It is most important to identify members to serve on teams that will focus on specific subsystems of the product, and a team leader for each. Typically the team leaders are drawn from within the product's own organization. It is best if everyone is physically present in the same room. We have actually had very successful trans-Atlantic brainstorming meetings with three locations linked by speakerphone, but we do not feel that this practice is as effective as having everyone together.

You also must have a prepared "script" for the meeting, because you will be trying to:

- inform everybody about the product and its importance;
- motivate the people to put their effort into the Target Costing effort;
- make clear that there is a market-driven rational basis for the challenging cost targets.

106  Chapter 7

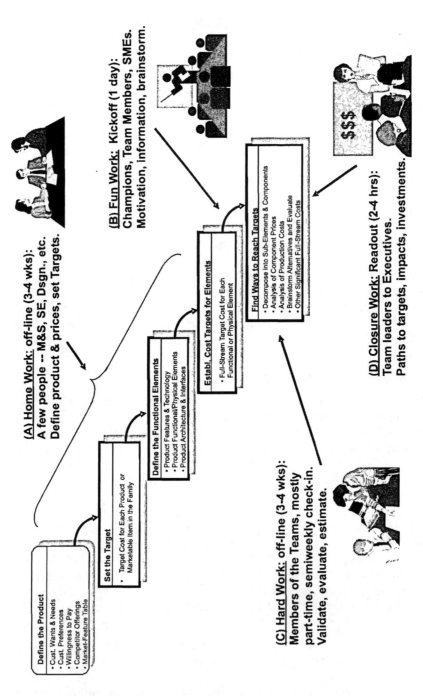

**Figure 7.3**  Information gathered at each point, and the time and effort required, in a Target Costing project.

## Putting Target Costing into Practice

| | | | |
|---|---|---|---|
| 1. | 9:00 am | Target-Costing Goals & Objectives | Executive Champion |
| 2. | 9:20 am | Overview of the Target Costing Process | Target Costing Person (you) |
| 3. | 9:40 am | Business Objectives for the Product | Product Manager |
| 4. | 9:55 am | Customer Needs (Market Feature) | Product Manager |
| 5. | 10:15 am | Competitive Situation | Competitive Analyst |
| 6. | 10:35 am | Product Architecture & System Elements | Architect |
| 7. | 11:00 am | Overall Target Price & Cost | Product Manager |
| 8. | 11:20 am | Baseline Costs & Cost Targets | Target Costing Person (you) |
| 9. | 11:45 am | Members of Subsystem Teams | Target Costing Person (you) |
| 10. | 12:00 pm | *Lunch* | |
| 11. | 1:00 pm | Initial Brainstorming (all) | Led by TC person and Prod. Mgr. |
| 12. | 2:15 pm | Breakout Brainstorming | Teams |
| 13. | 4:00 pm | Readouts | Teams |
| 14. | 4:15 pm | Next Steps (through Final Readout) | Target Costing Person (you) |
| 15. | 5:00pm | *Adjourn* | |

**Figure 7.4** Suggested agenda for a one-day brainstorming meeting.

The first half of the brainstorming meeting typically takes 2–3 hours, and is intended to get everyone up to speed, with them hearing the same consistent story at the same time. After that, the brainstorming can begin. We find it best if there is a brainstorming session involving the entire group. It is often surprising how many useful ideas in any area are generated by people whose primary responsibility is in a different area. The cross-fertilization is very useful. Then subsystem teams should break out to do their individual brainstorming, and report back their additional ideas. All ideas should be captured and recorded. Fig. 7.4 shows a typical agenda and list of speakers for an internal meeting.

In some cases you may choose to conduct your Target Costing meeting as described and then have focused meetings with selected suppliers as described in Chapter 5. Or you may choose to involve key suppliers in the Target Costing meeting. Suppliers are invaluable participants because they bring additional subject matter experts to address the challenges. They also can commit additional resources to help implement the brainstormed ideas. When you engage suppliers, you will have additional considerations, especially if you invite potential competitors to the same meeting. To create the appropriate environment frequent, early communication of your company's needs and intentions are required. Detailed nondisclosure agreements are a must to ensure the appropriate exchange of ideas. Some ideas are suitable for all participants and encourage collaboration. Others will be proprietary for one company or another and can be discussed in separate breakout rooms and handled appropriately.

## FINDING PATHS

Typically you will come out of a brainstorming meeting with 100–150 good ideas for taking cost out of the product. We have seen as many as 250. The next step involves considerable effort and can take from 3 to 6 weeks. Fortunately, this work can be distributed across many of the people on the subsystem teams and so each person may have to work only part-time to help bring this phase to its conclusion. First, you must organize the ideas. For example, some of the ideas may be redundant, or essentially identical and just expressed in different words. Some ideas may be *clearly* impractical, infeasible, or incompatible with clearly defined market needs. So after an early triage, you must put the remaining ideas into categories and assign the ideas to the teams. You may also have ideas that need to be investigated by individuals such as product managers, sales people, and so on. Such an idea might be one that says, "Let's eliminate this feature because it's expensive and it doesn't look like many customers are willing to pay for it."

Then the teams need to evaluate each and every idea. A team should first determine, "If we implement this idea (no constraints), how much money will it save?" This gives the worth of the idea. Only then should the idea be checked for technical feasibility or compatibility with the rest of the system. Assuming it is valuable and practical, the team should then estimate the amount of effort (time, people, resources, cost) to implement the idea. Note that many ideas can be implemented at *zero* incremental effort; they just require a change in direction, a change in emphasis, or a different choice.

As the Target Costing person, you should keep a running record of the outcome of every idea. Some possible outcomes could be "Drop it," "Hold it for a later release of the product," or "Implement it." You should organize regular short (30-minute) meetings of the team leaders to check on progress, resolve incompatibilities, and do tradeoffs. It may be possible to take a large amount of cost out of one subsystem by doing something that slightly increases the cost of another subsystem. Remember, while the teams each have individual targets that they are trying to reach, they are *not* in competition with each other. The important thing is that they collectively reach the overall target cost of the complete product.

## THE READOUT

Everyone should be aware that they are driving toward implementing a clear set of ideas that will bring the product within its cost target goals. There should be a scheduled readout meeting with the champions and other senior managers associated with the product. It is best if this date is set even before the brainstorming meeting, so that everyone knows that they are working toward a very important point in time. The readout meeting can be short, 1 to 2 hours. Most of the readout should be done by the product manager and the team leaders. It is essential to convey:

- the goals of the Target Costing project;
- the targets that need to be reached, and the initial gaps;
- the set of ideas that, if implemented, will bring the product to meet its target cost;
- the subset of ideas that can be implemented immediatly at zero incremental effort;
- the effort required to implement each of the others (time, people, resources, cost).

## ACHIEVING AND MAINTAINING TARGET COSTS

After the ideas are identified, the team and management will be eager to realize the benefits from the ideas generated. Follow-through is the biggest challenge you will face. Although some ideas can be implemented immediately without a significant investment of time or resources, others will be much more difficult. It is important to adequately document the complex ideas. You should show what the assumptions were, how they were validated, what the dependencies are, and some sensitivity analysis of the high-risk factors. Each idea needs a mini-business case to show what the investment and payback should be. This added discipline is crucial because the idea must stand on its own since team membership and other factors like customer demand may change. Months later when challenges arise, the rationale for the idea may be tested and idea originators may not be available to analyze the idea in the context of the new conditions. In these situations, such documentation helps the responsible team make an informed decision. Although this requires some additional documentation at the time of idea generation, much of the important information is available when project approval decisions are made. With the internet, hypertext links to supporting documentation can be provided for all projects generated from the list of ideas. This is especially convenient for globally distributed teams because it allows access to the information any time of day or night.

In Chapter 6 we talked extensively about templates for tracking progress toward achieving and maintaining target costs. Wide visibility of the gap and progress toward closing it is essential. For the plans, and thus follow-through, to be successful, there needs to be a regular review meeting. The team should be motivated to prepare for and participate in these meetings by the desire to actually close the target cost gap, rather than just talk about it. However, you will generate excitement if each participant has something to contribute and something to gain from the review meeting. Design each meeting so the participants can share their progress toward closing the target cost gap. At the begining of the meeting, provide an overview of major issues and dependencies for the entire portfolio. Also be sure to include a special topic that highlights a new opportunity, technique, or a change in a customer's or competitor's situation that has implications for the industry experience curves. These helpful hints will make long review meetings much more interesting. Short presentations of the standard templates will make it easy for your cross-functional team to see what has changed and keep everyone alert.

## CHECKLISTS

The following pages contain a checklist for each major step in the Target Costing process. We have found them to be very helpful, so we offer them for your use. You will notice reference to "peer reviews"; we have found it beneficial to periodically review a Target Costing project's progress with other practitioners who are not specifically involved with the project.

Checklist 1: Define the Product
- Get executive "buy-in"—identify champion(s).
- Identify preliminary Target Costing staffing needs.
- Identify business goals and metrics.
- Identify customers and addressable markets.
- Identify competitors.
- Analyze competitive offerings.
- Complete competitive analysis and reverse engineering (CA/RE).
- Identify customer wants and needs.
- Determine willingness to pay.
- Establish customer weightings (importance of features).
- Develop market-feature table.
- Complete a peer review.

Checklist 2: Set the Target
- Identify the scope of costs to be included.
- Establish target price (experience curve) and target profit.
- Establish target cost for the overall product.
- Establish cost targets for subsystems (value engineering), sub-subsystems, etc.
- Establish cost targets for other elements (e.g., installation, transportation, etc.).
- Establish business performance targets (e.g., shipping performance, etc.).
- Identify optimum provisioning locations (supply-chain analysis).
- Identify most competitive and most likely supply chains.
- Review and get agreement on full-stream targets (ROIC, NPV, etc.) with business entity.
- Complete a peer review.

## Putting Target Costing into Practice

Checklist 3: Find Paths to the Targets
- Establish Target Costing teams (include non-business-entity subject-matter experts).
- Plan the brainstorming meeting.
- Complete a pre-brainstorming peer review.
- Hold the brainstorming meeting.
- Assign the ideas to teams.
- Validate, evaluate, and estimate the ideas.
- Identify the final recommendations.
- Complete a pre-readout peer review.
- Hold the readout meeting.

Checklist 4: Achieve the Targets
- Quantify the required management support.
- Ensure that the initiatives are funded.
- Ensure that people are assigned to implement the recommendations.
- Project plans for R&D, Purchasing, and Manufacturing initiatives.
- Revalidate assumptions and market competitive intelligence.
- Fit the initiatives into an overall strategy.

Checklist 5: Maintain Competitive Costs
- Quantify gaps and plans to gain management support.
- Engage the Finance organization to report results at product level.
- Continue ongoing competitive analysis and reverse engineering.
- Ensure that ongoing projects are adequately staffed and funded.
- Ensure active involvement of Purchasing to insure material costs stay on industry curves.
- Use ongoing reporting to highlight issues early.

## SUMMARY

In this chapter we have provided some practical advice that you will find useful as you begin to put Target Costing into practice, and as you conduct your first few Target Costing projects. Above all, you must insure that other participants are enrolled, convinced that their efforts will lead to good business results, and understand that the results will reflect well upon them. Also, remember that every project is a little different, and you must emphasize some things more than others. Above all, you must be flexible and adaptable.

# 8
## Some Case Histories

**CASE HISTORIES**

In this chapter we offer a number of case histories from our own experience. They will illustrate some of the nuts-and-bolts details of what has to be done when putting Target Costing into practice. They also illustrate the maxim that "every project is different"—depending on circumstances, some issues need more emphasis than others. The reader will notice that, in many cases, actual numerical data is hidden or disguised. We did this to avoid disclosing proprietary information. Nevertheless, we do not believe that this will dietract from the lessons learned from studying these case histories.

We present three case histories, each illustrating important aspects of Target Costing:

- **Cellular Base Stations**—a case where the market-feature table defined the architecture, and hence the cost, and how a credible target cost inspired a team to beat the competition.
- **Outdoor Electronics Cabinets**—a case where Conjoint Analysis and reverse engineering, applied in a market of a few well-known business customers, revitalized a slumping business.
- **Optical Interface Shelf**—a case where the market feature table resolved the opposing views of Marketing and R&D. It also shows how the same team appropriately applied market and competitive analysis to halt what initially looked like a promising product.

## CELLULAR BASE STATIONS

This case history shows the importance of "Define the Product" and of discerning what customers are really willing to pay. It was one of the first projects that we worked on (in 1997) that involved a product that was large, complex and, configurable, that is, it could be configured to adapt to a wide variety of customer applications. The product was a cellular base station—the telecommunications equipment that resides at the base of the antennas used in cellular telephony. They handle the radiofrequency (RF) transmission and reception to and from mobile cell phones, and handle the traffic flow to the rest of the telephone network. They also manage the hand-off from one base station to the next as the customer moves about the region.

In this case the company already had a family of cellular base stations and other equipment that handled this particular cellular technology, known as CDMA (code-division multiple access), which is a way of using the RF spectrum to accommodate high numbers of simultaneous conversations. The first-generation product was very successful, with high market share. And it was undergoing cost-reduction efforts. But technology advances and anticipated new competitors' offerings dictated that it was time to introduce a next-generation product. The new product was to have enhanced features, take less space, use less power, etc.—and cost less. In fact, early studies of a preliminary architecture and design suggested that costs—and prices—could be reduced by a factor of two by the time the product was to be introduced into the marketplace.

In CDMA cellular base stations the number of "RF carriers" installed affects the number of subscribers that can be served by the station. RF carriers can be thought of as being similar to "channels" or frequencies in normal radio or TV communications. And the customers who buys cellular systems (the service providers) think in terms of the cost incurred to serve each subscriber. Fig. 8.1 plots base-station costs (in dollars per subscriber) as a function of the number of carriers in the base station. The cost of first-generation ("previous") product—which could accommodate only two carriers—before and after cost reductions is shown as the solid and dashed lines, respectively, in the upper left of the figure. The original business-case plan for the new second-generation product ("original business case") is the lower solid line. This looks like a substantial improvement in cost, so one might ask, "Was there really a need for Target Costing?" As will be seen, there was.

First we started working on "Define the Product." We spent a lot of time with the Product Managers and the domestic and overseas customer representatives, getting a better picture of what customers wanted, needed, and were willing to pay for. We also gained a good picture of competitors' product offerings, both existing and promised. From this we developed breakdowns in terms of market size, market geographic distribution, types of installation, ratio of startups to growth systems, and so on.

## Some Case Histories

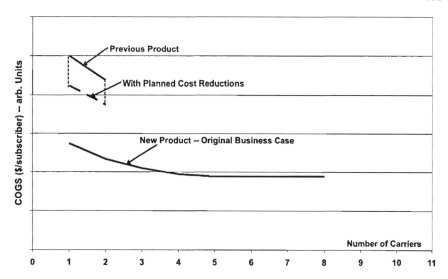

**Figure 8.1** Starting point for cellular base stations.

Fig. 8.2 indicates some of the kinds of information we gathered. For proprietary reasons we must disguise the actual data. But the figure indicates the kind of information that was important, if not the actual values—size of markets, geographic distribution of customers, indoor versus outdoor installations, startup versus growth systems, application environment (urban, rural, etc.), and system size. One of the most significant things that we learned was the size of the system (in terms of number of carriers) that customers would be ordering. It became clear that service providers wanted to buy inexpensive systems with one or two carriers, to get on the air at minimum cost. Then, as their number of subscribers and traffic increased, they would grow eventually to large size by adding carriers, one or two, at a time. In fact, all initial sales would be for one, two or three carriers. It turned out that the initial architecture resulted in a system that was cost-optimized at six to eight carriers, where there would be no sales. This discovery led to a realization that the costs had to be much lower at the one-, two- and three-carrier configurations and led to some breakthroughs in architecture. We used this and much other information to develop a one-page market-feature table for the various market segments. This provided further clarity about what functions and features were truly important to the customers.

Next we turned to "Set the Target." We developed a preliminary experience curve that suggested that cellular base stations were on a 73% curve, that is, prices were declining by 27% for every doubling in the number of cumulative worldwide cell-phone subscribers. The number of subscribers is the correct metric here, because that is the factor that drives the market's technical and economic experience. And we expressed

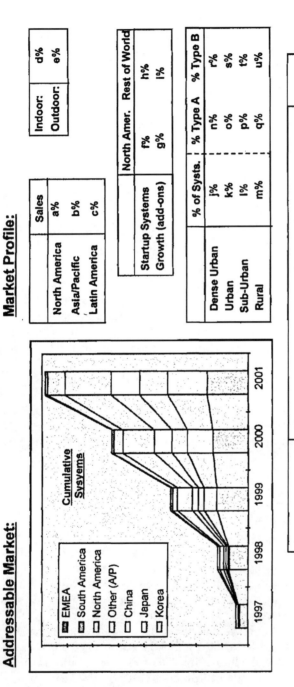

**Figure 8.2** Some information from "Define the Product."

# Some Case Histories

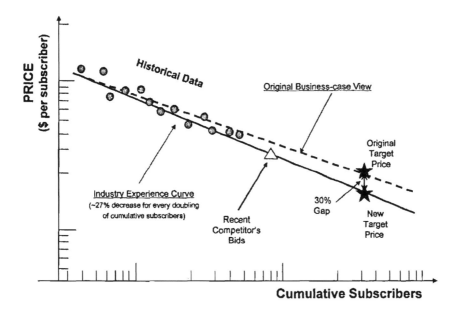

**Figure 8.3** Experience curve (price) for cellular base stations.

the price in terms of dollars per subscriber, which takes into account the price of a system and the number of subscribers it can support. The curves are shown in Fig. 8.3. Note that we were working on this project in 1997 and we were projecting the likely prices for late 1998, when the product was to be introduced. Our projections were confirmed in real time, when the company was involved in some competitive public bids for a large network. The winning bids fell right on the projected price experience curve, giving us confidence that we had to aim for lower prices for the new system than originally planned.

After the required gross margins were factored in, it was clear that the new system would have to be developed to meet a target cost about 30% lower than had been originally planned. This can be seen in Fig. 8.4, where we show the required costs as a function of the size of the system. Note that these costs apply to the time of the product's planned introduction.

At this point, we had determined the customers' wants and needs and put them into a market-feature table. And we had established price and COGS for the overall system. Next we partitioned the system into sections and set cost subtargets for each section. This is shown in Fig. 8.5; the major sections were the RF electronics, the digital control electronics, the "equipment" (including cabinets, power supplies, alarm boards, etc.), and manufacturing (including both circuit-pack assembly and test, and final-system assembly and test).

We formed cross-functional teams and gave them targets for the subsystems. We also set aside a "kitty" of about 5% to relieve especially

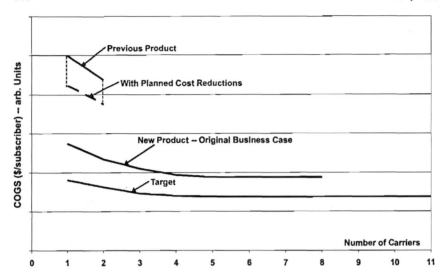

**Figure 8.4** The market-based target cost for cellular base stations.

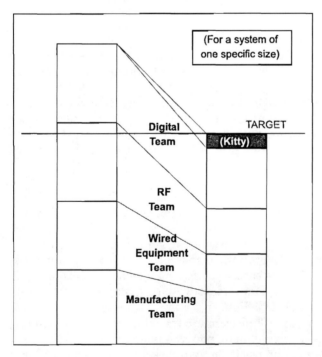

**Figure 8.5** Subdividing the target cost.

difficult cases. It is not shown in Fig. 8.5, but we actually subdivided each system's cost target into cost targets for each circuit pack or other element that comprised the subsystem. The teams were made up principally of designers of the elements in question. But we also brought in people from other parts of the company with relevant technical, engineering, and cost experience. We also had a steering team that included product managers, architects, systems engineers, installation engineers, and more besides.

We gathered everyone, about 50 people, into a single place and had an all-day "kickoff meeting." In the first half of the meeting the executives and product managers explained the marketplace rationale for the product's features and price/cost targets. This gave everyone a united view of the motivation for the Target Costing project. Then we had two brainstorming sessions: the first included everyone in the room and the second involved the subsystem teams working individually. By the end of the day we had over 100 ideas for reducing the cost of the system.

It took us about two and a half months to get to this point, and the next phase of the work took about a month. The teams had to validate the ideas (ensure that they were feasible, and compatible with everything else), evaluate the savings that the ideas would provide, and estimate the amount of effort (time, cost, personnel) to implement the ideas. During this phase, we arranged for the teams to touch base frequently to report progress and to do tradeoffs. For example, it's OK to add $200 to the cost of a digital controller if it removes $650 from the cost of the FR amplifier. At the end of this phase the teams made a "readout" to the executives, explaining the targets, the proposed best set of paths to achieve the targets, the effort required to achieve them, and the risks and dependencies. The principal improvements in cost arose from architecture changes, new RF amplifiers, improved RF technology and packaging, simplified hardware and cabling and power architecture, more aggressive integration and densification, and lower testing costs. Fig. 8.6 shows the final result (lower irregular line with data points), which was accepted and implemented.

It is worth noting that in the 12–15 months that followed before the product was introduced into the marketplace, the design team was able to stay within ±1% of the result presented at the readout. The product was introduced at a competitive cost and became a successful, dominant player in the marketplace.

## OUTDOOR ELECTRONICS CABINETS

This case study (which was already mentioned in Chapter 1) shows the use and impact of Conjoint Analysis in Target Costing.

A telecommunications equipment manufacturer also made the outdoor electronics cabinets that housed the equipment. The cabinets were sometimes sold bundled with the equipment and sometimes as stand-alone products. The company found itself losing market share in

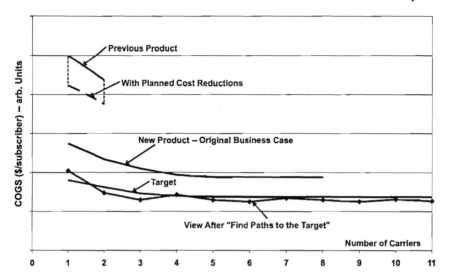

**Figure 8.6** Target Costing results for cellular base stations.

both the equipment and the cabinets, and part of the cause was the high price (due to high cost) of the cabinets. They decided to redesign the product line, including the cabinets. Although part of a larger project, this case study focuses on the cabinet product family redesign.

In this application, the customers were well known because the company's current customers covered most of the addressable market for the product. Even though the customers were well known, the fact that they had different needs led to debates in the electronics cabinet company that needed resolution before design could begin. Even the stakeholders within each customer company had different needs, further complicating the understanding of what the customer really wanted.

To assess the situation, the cross-functional team consisting of Product Management, Sales, Design, and Manufacturing created a market-feature table. As the chart was assembled, it became clear that the different subject-matter experts had opposing views of what the customer wanted. Fig. 8.7 shows an early classification of the features for the OEM (original equipment manufacturer) market.

To settle the debate, competitors' cabinets were purchased for reverse engineering. They ranged from low-cost cabinets for limited applications to high-end, fully featured cabinets suitable for harsh environments. From competitive and reverse engineering analysis, detailed comparisons were developed at the subsystem level. Fig. 8.8 shows the list of features created for one subsystem, the AC power panel. The company's (Company A in the figure) current solution had every possible option as part of the standard product for most subsystems, whereas some competitors only offered products to provide basic functions. When looking at the whole cabinet compared to what competitors offered, it became evident the old product

# Some Case Histories

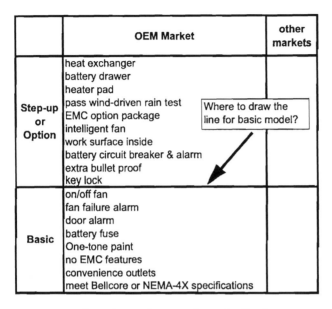

**Figure 8.7** Question—where to draw the line?

| AC Power Panel Features | Price Leader | Company A |
|---|---|---|
| UL listed | | X |
| commercial load center panel | X | |
| custom-made box | | X |
| color coordinated | | X |
| AC surge protected | | X |
| exterior generator inlet | | X |
| interior generator inlet | X | |
| convenience outlets standard | | X |
| exterior transfer switch | | X |

**Figure 8.8** AC power panel features.

line was a complex, fully featured, high-cost solution, but some customers were satisfied with low-cost, basic functionality.

Realizing that the company's customers needed advanced features as options, the challenge became deciding where to draw the line between basic features that every customer needed and step-up features that could be added for those willing to pay for them. The project goals became to redesign the product family to restore high return on sales by providing low-cost, basic functionality outdoor electronics cabinets with options that would meet the needs of fully featured customers. Subsequently, the tactical goals included a detailed market requirements analysis. In short, the goal was to develop a "southwest corner" cabinet as depicted on a chart of price versus features as shown in Fig. 8.9.

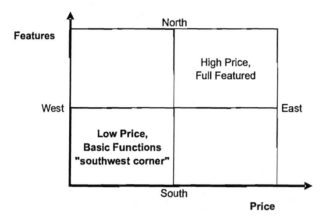

**Figure 8.9** Goal: develop a "southwest corner" cabinet.

Initially, the team listed the attributes that influence the purchase decision. It included things like price, type of equipment the cabinet would support, door style, approach for future growth, and order interval. Then they determined the key design tradeoffs. In addition to these key customer attributes, designers brought up questions like where to put the protection blocks and what battery technology to provide. The manufacturing team brought up color. The reason was that the old design had two colors. If there were only one color, there would be savings in paint-line changeovers, assembly steps, and inventory management. Sales brought up delivery. For each attribute there were several valid options, but the team was not sure which combination would best satisfy the diverse customer base. The team determined viable costs and likely prices for each possibility. The list shown in Fig. 8.10 summarizes the tradeoff decisions that influence the cabinet purchase decisions. This became the basis for a Conjoint Analysis survey created using the ACA$^{TM}$ program from Sawtooth Software.

The team developed the purchase scenario of Fig. 8.11 and set up meetings with key stakeholders in customer companies. There were seven major companies in this market and they were all long-term customers of the cabinet manufacturer. This greatly simplified collecting the customer data, and allowed the team to use data from a few respondents to represent large portions of the addressable market. In most cases, 200 respondents are required to get a statistically significant sample size. Here, fewer than 100 covered 70% of the total addressable market. Many of the individuals who made the purchase decision were well known and visited regularly as part of ongoing business. They were delighted to extended a regularly scheduled meeting to include a discussion of the next-generation cabinet. The one-hour meeting consisted of a presentation on the emerging new design, a paper survey to collect demographic data and open-ended comments, and the 20-minue computer-driven

# Develop Critical Attributes & Levels that impact design or purchase decisions

- **Color**
  - No exterior paint - bare aluminum ($300 discount)
  - Light gray powder painted exterior (standard)
  - Beige powder painted exterior (standard)
  - Custom color exterior: brown, blue, or other for $450
- **Equipment**
  - Mixed vendor A & vendor B equipment in same cabinet
  - All vendor A equipment in cabinet
- **Growth Approach**
  - Pre-wired for full capacity
  - Wired for current need now; purchase slide-in bay & additional material kit later
  - Separate, additional cabinet later
  - Extension cabinet later
- **Door Styles**
  - Swing equipment chamber doors; liftoff end chamber doors
  - Swing equipment chamber doors & swing end chambers for $250
  - Overhead equipment chamber doors & swing end chamber doors for $2,000
- **Battery Chamber**
  - Battery (3 yr. life) on shelf, liftoff door accessible
  - Battery (3 yr. life) in pull-out drawers for $600
  - Battery (7 yr. life) in conditioned compartment for $1,200
- **Auxiliary Power**
  - 8 hours battery operation for low traffic rates
  - 8 hours battery operation for low traffic rates and gas generator inlet for $300
  - $6,000 module for long battery reserve or generator
- **Protection Blocks**
  - Protection blocks in end chamber for more equipment space
  - Protection blocks in equipment chamber
- **Order Interval**
  - 4 weeks from receipt of order to receipt of cabinet ($600 discount)
  - 3 weeks from receipt of order to receipt of cabinet ($200 discount)
  - 2 weeks from receipt of order to receipt of cabinet
- **Delivery**
  - Separate shipments of cabinet, batteries, & circuit packs ($600 discount)
  - Whole order delivery (components arrive together)
- **Price**
  - $19,000
  - $20,000
  - $21,000
  - $22,000
  - $23,000

**Figure 8.10** Tradeoff decisions that influence the outdoor electronics cabinet purchase decisions.

> *Scenario:* Assume for the purposes of this survey that you have an outside electronics cabinet application requiring 768 telephone lines. Eventually, you will need 1,500 lines, but you are not sure when. You plan to buy a small electronics cabinet for $21,000 that meets your initial needs and has space for growth. Included in the price are the cabinet, heat exchanger cooling, equipment fans, AC power panel, protection blocks, and battery cables. Items not included are telephony equipment and batteries.
>
> Prices and options are for hypothetical products and do not describe an actual product or pricing. Your responses will be used to design a cabinet that meets your needs. Some options will be desirable and others undesirable. This will take 20 minutes. Thank you for your participation in the new design!

**Figure 8.11** Electronics cabinet survey purchase scenario presented to customers.

Conjoint Analysis survey. Their participation paid off a year later when the new cabinet family was introduced reflecting their input.

The results for each attribute were summarized graphically. Actual buyer utilities are left off the charts in Fig. 8.12, but the shapes reveal some subtleties that were not known before collecting the customer input. On the question of color, the responses showed that customers prefer beige even if there is a discount for no paint. Although they were buying brown cabinets and sometimes said they wanted other colors, the survey revealed they did not want to pay what it cost. The quantification of individual responses showed no significant difference between gray or beige. Interestingly, gray was finally selected because one major customer strongly favored gray and the others did not care! Because the discussion with customers had changed to which color rather than two, manufacturing was able to reduce processing time and inventory, resulting in savings. On price, the curve reveals that lower price is preferred. Here again, the ability to look at curves by company provided new insights. It was learned that a small saving was desirable, but if prices were too low, customers would be concerned that something important was missing. We also learned that one customer would accept a small price increase. The data revealed other insights that the sales team used later when selling the new cabinet because they could highlight the desirable features that individual customers had selected in the survey.

Armed with insights of the market, the team was ready to brainstorm solutions to meet the target cost. Combining inputs from the competitive tear-downs, customer feature and price requirements, and gross margin goals, they set the target cost based on market drivers. To brainstorm solutions to achieve the target, the core team called in expertise beyond their small team and even beyond their business unit. In a day-long meeting, the team presented the challenge to the wider cross-functional team. The whole group watched as the core team took each subsystem and further broke it down to the component level. Interfaces between the subsystems were well documented so everyone could see which issues extended across subsystems. For instance, the subsystem for

## Some Case Histories

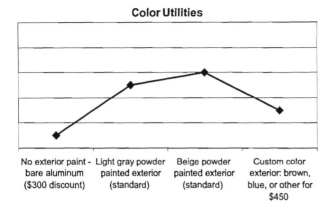

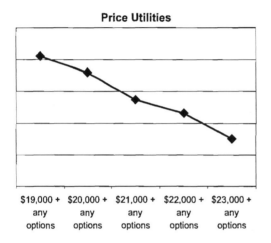

**Figure 8.12** Buyer utility data from outdoor electronics cabinet surveys.

battery cables, venting, and heating, depicted in Fig. 8.13, led to questions about wire size. At this level, the outside experts could contribute without regard for any perceived constraints of the cabinet business. Since the team had studied the differences between existing cabinets, someone asked why 6- and 8-gauge wire was being used in applications where other companies used lighter 10- and 12-gauge wire. The reply was that those companies were only designing for one fault condition and the heavier gauge wire would cover multiple fault conditions occurring at the same time. Had this ever happened before? No. So a switch was made to 10- and 12-gauge wire. The net effect of this change was small, but the questioning continued. What about the interfaces, that is, the connectors on the ends of the wire? It happened that they had been a source of frustration as they were a custom part and had caused delivery problems in the past. As a result of the wire gauge change, a standard connector was selected resulting in significant full-stream savings.

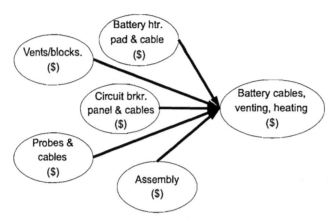

**Figure 8.13** Battery cables, venting, and heating subsystem for outdoor electronics cabinet.

After the meeting, the core team considered all the concepts from the brainstorming session. They clarified concepts, validated assumptions, conducted experiments, obtained quotes, and made the necessary trade-offs to agree on a path to achieve the target costs before proceeding into the design phase. This approach allowed the core team to benefit from the wide expertise of the entire company, without a large investment of time by the outsiders.

One year later, a new family of cabinets was introduced that met the customer needs and beat the competition in critical price and performance metrics. By applying elements of Target Costing, especially competitive tear-downs and market analysis, the team redesigned the cabinets and the manufacturing process to achieve a 30% cost saving over the previous designs. With its new product line, the firm's revitalized cabinets business was growing again, as indicated in Fig. 8.14. Shortly after introducing the new product line, the business returned to profitability and within two years revenue doubled.

## OPTICAL INTERFACE UNIT (OIU) FOR TELECOM SWITCH

This case study emphasizes the value of the market-feature table, value analysis, and a broad cross-functional team. In the mid-1990s, optical technology was providing huge efficiencies in telecommunications networks. In 1998, a telecommunications equipment manufacturer was designing a new optoelectronics shelf that would switch at narrowband levels (DS0) but interface to optical broadband (OC-3). DS0 and OC-3 are transmission rates of 1.5 Mbit/sec and 155 Mbit/sec, respectively. Product management, design, and manufacturing were in three different states. The lack of co-location had led to an "over the wall" culture. Typically, product management would develop design requirements based on customer discussions, and then the design team would develop design

# Some Case Histories

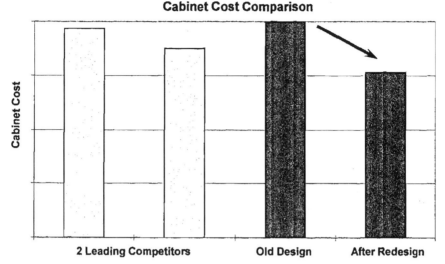

**Figure 8.14** Restoring electronics cabinets to profitability with Target Costing.

proposals for product management approval. Once a prototype was designed, there would be a manufacturing design review. Then, after a few iterations, the design was transferred to manufacturing. The OIU team broke this pattern.

The first step was to define the product using the market-feature table. There were two major markets, North American (NA) and international. Product management created an initial list of features in each category. Fortunately, with minimal changes, agreement was reached. Although a simple chart, it pointed out that the only differences between the NA and international versions were in the communication protocol standards used. Design was already investigating chip sets that could either support both or be swapped depending on the protocol. Therefore the basic product was simpler than originally thought. Merely organizing the information and publicizing it for all stakeholders simplified the design challenge. It also set priorities by focusing the team on the basic requirements and deferring some premium feature development. The resulting market-feature table for the NA market is shown in Fig. 8.15. Agreement on this simple chart was the catalyst to changing the "over the wall" culture and establishing a cross-functional team with a common goal.

Once the product was defined, the next step was to set the target cost. Because the OIU provided new capabilities, setting the target cost presented special challenges. There was a good history for electrical interfaces, but no optical interface units for this application existed yet. And, unlike the cabinet case study, there were no competitive products in the market for reverse engineering. Two steps were followed to

| Need Level | North America Market Segment Features |
|---|---|
| Premium | OC-3 ADM functionality<br>OC-12 ADM functionality |
| Step-up | "Remoteable" I/O shelves<br>STM-1 in same shelf<br>Duplicated Unit Controllers |
| Basic | Multi-vendor intraoffice point-to-point OC-3 interface<br>1x1 line protection (redundancy for critical data)<br>1x1 equipment protection (redundancy for power, etc.)<br>SONET overhead terminations<br>DeMUX to DS0 level<br>VT-based OC-3<br>DS1 & OC-3 alarm levels<br>Performance monitoring<br>Small footprint; fit in current bay |

**Figure 8.15** "Basic" product requirements for proposed optical interface unit.

compensate. First, data for electrical interfaces was gathered and then converted to the basic unit, DS1, even though the requirement was for higher-capacity optical, OC-3 interfaces. Thus the OC-3 capacity was represented by equivalent DS1's. This gave a fundamental metric that could be used for comparisons across products. Customers were interested in *capacity*, hence the development of high-capacity optical interfaces. Fortunately, other products had recently gone through a similar technology shift for greater capacity. Therefore, once the experience curve was derived for DS1 equivalents with electrical interfaces, a density factor converting from electrical to optical technology was applied. The density factor was derived from the optical/electrical transport technology ratio seen in the other products. The target cost was developed at the fundamental level and then scaled back up for the capacity of the OC-3 interface and the basic model consisting of 168 DS1 equivalents per shelf. The baseline (starting-point) cost of the product was $158/DS1 equivalent. The target cost derived for the end of 1999 was $113/DS1 equivalent and $104/DS1 for the end of 2000. Using linear interpolation, the OIU target for introduction in March 2000 was $110/DS1 equivalent. For the proposed 168 I/O OIU shelf, the target became $18,480. The target cost derivation is summarized in Fig. 8.16.

The next step was to develop an integrated view of the relative importance—to the customers—of the key functional needs or characteristics of the product. Product Management provided the importance table shown in Fig. 8.17. The next step after that related the major subsystems to the customer-driven features with a function–subsystem matrix, as shown in Fig. 8.18. This process was completed *and agreed to* by a cross-functional team in only 3 hours.

## Some Case Histories

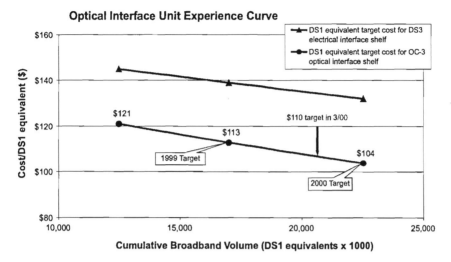

**Figure 8.16** Target costs (per DS1 equivalent) for the optical interface units.

| Customer Need & Importance (from survey) || Detailed Functional Need (allocated by product management) ||
|---|---|---|---|
| Need | Importance | Detailed Functional Need | Importance |
| OC-3 Termination | 35% | Multi-vendor intraoffice point-to-point OC-3 interface | 10% |
| | | 1x1 line protection (redundancy for critical data) | 10% |
| | | 1x1 equipment protection (redundancy for power, etc.) | 10% |
| | | SONET overhead terminations | 5% |
| Mux Function | 20% | DeMUX to DS0 level | 10% |
| | | VT-based OC-3 | 10% |
| Processing | 10% | DS1 & OC-3 alarm levels | 5% |
| | | Performance monitoring | 5% |
| Size | 35% | Small footprint; fit in current bay | 35% |

**Figure 8.17** Importance of OIU features to customers.

Looking at the subsystem gaps, some meet the target and other exceed it. Reallocating the subsystem targets would not close the total gap. Looking at the gaps highlighted potential issues with the DS0 card, backplane and cage, and power module card as shown in the value graph of Fig. 8.19.

After reviewing the relevant competitive analysis and the initial (baseline) design, the cross-functional brainstorming team examined each circuit pack, carefully highlighting the major cost drivers. The cross-functional team brainstormed to generate ideas that could lead to lower costs. Dozens of ideas were generated, evaluated, and eventually implemented over the next few months as depicted in Fig. 8.20.

| Customer need & importance | | OC-3 Termination 35% | | | MUX Function 20% | | Processing 10% | | Size 35% | | | | |
|---|---|---|---|---|---|---|---|---|---|---|---|---|---|
| Functional importance | Intra-office OC-3 interface | 1x1 line protection | 1x1 equipment protection | SONET overhead terminations | DeMUX to DS0 level | VT-based OC-3 | Alarms | Performance monitoring | Compact Size | | | | |
| Sub-system | 10% | 10% | 10% | 5% | 10% | 10% | 5% | 5% | 35% | Target | Baseline | % gap | $ gap |
| DS0 card (x8) | | 20% | 35% | 70% | 40% | 25% | 40% | 100% | 35% | $ 6,422 | $ 14,593 | 127% | $ 8,171 |
| Backplane & cage | 10% | 10% | 10% | | | 5% | | | 10% | $ 1,294 | $ 2,874 | 122% | $ 1,580 |
| Power card (x2) | | | 10% | | | 5% | | | 10% | $ 924 | $ 1,999 | 116% | $ 1,075 |
| Formatter card (x2) | | | 10% | 10% | | 10% | | | 10% | $ 1,109 | $ 1,859 | 68% | $ 750 |
| Fuse card (x2) | | | 10% | | | 5% | | | 10% | $ 924 | $ 713 | -23% | $ (211) |
| Cabling & misc. | 10% | 5% | 5% | | | | | | 5% | $ 601 | $ 143 | -76% | $ (458) |
| Optical card (x4) | 80% | 50% | 15% | 20% | 30% | 20% | 20% | | 15% | $ 4,943 | $ 3,683 | -25% | $ (1,260) |
| Unit Controller | | 20% | 5% | | 30% | 30% | 40% | | 5% | $ 2,264 | $ 790 | -65% | $ (1,474) |
| Value-based cost | $ 1,848 | $ 1,848 | $ 1,848 | $ 924 | $ 1,848 | $ 1,848 | $ 924 | $ 924 | $ 6,468 | $ 18,480 | $ 26,654 | 44% | $ 8,174 |

**Figure 8.18** Feature–subsystem matrix for OIU with target cost broken out by subsystem and customer value.

# Some Case Histories

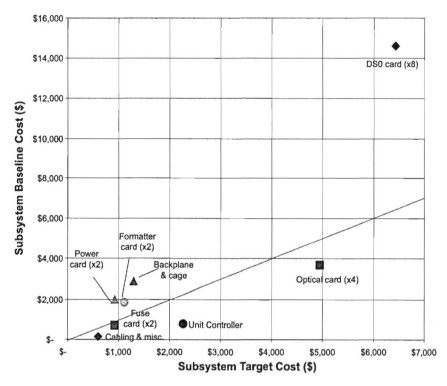

**Figure 8.19** Value graph for OIU, showing baseline costs versus target costs.

Immediately after the brainstorming session, component substitution recommendations provided by subject-matter experts outside the design team were used without significant analysis and few design changes. The large gap in the DS0 card subsystem led to questions about the protection scheme. Most of these cards were redundant backups duplicating the customers' critical data in case something went wrong. The most significant change was to the signal protection scheme. This required a software change. Initially there was reluctance because software changes often lead to long developments or, worse, bugs discovered later. In this case, further investigation revealed that the processors were fast enough, the software changes minimal, and thus an entire DS0 circuit pack was eliminated. With this circuit pack eliminated, attention turned to evaluating the power architecture. Power was consolidated in redundant power packs that took card slots in the shelf. By changing the height of the circuit pack and distributing power modules on the circuit packs, it was possible to double the capacity of a shelf. The customer's view of cost was actually at the system level, not the shelf. Therefore, eliminating a shelf in the bay would reduce overall cost. Or, stated another way, fewer shelves increased cost per shelf, but cost per

132                                                                                                  Chapter 8

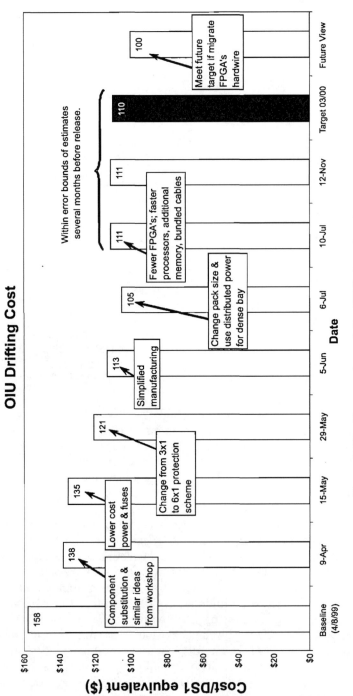

**Figure 8.20** Progress toward the OIU's target cost.

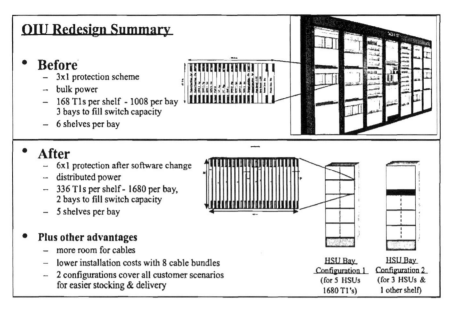

**Figure 8.21** Optical interface unit shelf design changes doubled the functional density and provided other benefits while meeting the target cost.

DS1 equivalent and total system cost declined. The pack-height change led to backplane and shelf changes that would have allowed the team to beat the target cost. However, they would have also led to a cable wiring challenge. Involving installation engineers and reviewing the future customer need led to bundled cables and a migration path that, although slightly more expensive initially, permitted a lower future cost. A few enhancements were added bringing the drifting cost up to the target cost. In the end, a hardware reduction of 30% was achieved, and additional savings were achieved in installation for a net impact of over 40%.

In the end, the new design did achieve the target cost because the product team had understood the customer requirements and competitive situation and leveraged the expertise of a larger cross-functional team. As a result, they had doubled the capacity of a shelf and simplified installation, as is shown in Fig. 8.21.

### Jumbo Telecom Switch

This continuation of the OIU case shows the use of Target Costing to help make strategic decisions about a proposed product.

Following the introduction of the optical interface shelf, the team embarked on a project to develop a high-capacity switch. Their initial switch design provided double the capacity of the existing switch. The market feature table's basic features are shown in the first column

|  | Relative Importance (%) | |
|---|---|---|
| Basic Product Requirements (top 8) | All Customers (equal weight) | Large Network Customers (78% of revenue) |
| Transmission Interface | 24% | 15% |
| Backward Compatibility | 12% | 12% |
| Management Interface | 12% | 12% |
| Reliability | 10% | 10% |
| **Capacity** | 9% | 18% |
| Delivery Interval | 9% | 9% |
| Density (DS1/ft$^2$) | 6% | 6% |
| Alarm Features | 6% | 6% |
| All Others (each 5% or less) | 12% | 12% |
| TOTAL | 100% | 100% |

**Figure 8.22** Relative importance of basic features in a market feature table to certain customer segments.

of Fig. 8.22. According to customer surveys, the first priority was the transmission interface, when each customer was given equal weighting. The switch customers consisted of both large and small network telephone companies. The large network companies represented 78% of the expected revenue for the new telecom switch. Among these large users, Capacity was the first priority, and the larger the capacity the better! Knowing this, the team focused on a much higher capacity Jumbo Switch.

During the design concept stage, a competitor's switch was introduced with four times the capacity of the existing switch. Reverse engineering revealed some alternative approaches and reinforced a challenging target cost. The goal became meeting this target cost while maintaining backward compatibility, the third highest priority. Although it was possible to grow the current switch with the existing technology to the capacity required, it could not be done without exceeding the target cost. Therefore this product development was terminated, saving $35 million. More importantly, this prevented bringing a product to market that would either not have satisfied the market need or met the target cost. A totally new product was developed later to provide the large capacity switch for the larger telephone companies. This illustrates the importance of defining the product and understanding and quantifying the customer requirements, even if it leads to not developing a product. Although stopping development of a popular product was challenging, it was certainly better than developing the wrong one for the changed market.

# 9

## Wrap-Up and Conclusion

This book and its simulation exercises began during the explosive economic growth of the late 1990s and were completed during the economic slowdown of 2002. During the growth years the importance of Target Costing was accepted for developing competitive products and was accepted and used by many companies as described in the text. Recently, the worldwide economy has been declining, and the importance of developing competitive products is even more important. A misstep in a new product can jeopardize the product and even the company. Renewed scrutiny of corporate financial reporting has also raised concerns about other corporate processes that affect the financial performance of any organization.

In this environment of declining growth, restatement of corporate earnings, and increased competition, product developers are under increased pressure to get it right the first time. There is increased emphasis to get products to market faster and without revision. Getting close to customers, fully quantifying customers' requirements and willingness to pay, and meeting corporate profitability are all part of the renewed emphasis on effective product development.

The role that Target Costing plays in product development during a shrinking economic environment is just as relevant, if not more relevant, than in a growing economic environment. Market-driven product design is therefore a tool for all seasons. Its relevance to successful product design and profitable product performance continues.

# Module A

Create a Business with a Strategy

**INTRODUCTION**

To illustrate the Target Costing concepts presented in the text, you are encouraged to work through the relevant module after completing each chapter. The Exercise is a fictitious product-design scenario. It was inspired by an article we read that used a wrist watch to illustrate key concepts of target costing [1]. We recommend that you alternate systematically between the chapters and the Exercise modules, because each module is intended to illustrate the concepts discussed in the chapters.

If you are in a classroom setting, you may form teams of 4–6 people and use the Exercise as a classroom exercise. The teams can each try to achieve the best financial result by the end of the Exercise. We should point out, however, that the teams are not really competing against each other—they are working against adaptive competitors in the study's hypothetical marketplace. Therefore, the teams are encouraged to share their strategies, data, and results at the end of each module.

The Exercise traces a hypothetical project to conceive, develop, and bring to market a next-generation product—a *wrist videophone*. It includes inputs for the Target Costing, as well as templates for organizing and graphing key information, setting cost targets, designing the product, finding paths to achieve the targets, and simulating the marketplace response. When done as a classroom exercise, the instructor may use the simulation module of ACA®, a conjoint analysis software package from Sawtooth Software, to provide a market share based on the parameters of the exercise. We provide in Module F a quick, back-of-the-envelope method to approximate your market share and compare it with others who do the exercise. Each module contains five parts; they are:

- Input and information that you get for free.
- Data and information that you "purchase."
- Activities that you do.
- Outputs that you generate.
- Discussions that you have (if you are in a classroom setting).

**INPUT: THE SCENARIO OF THIS EXERCISE**

**Background**

It's the year 2010. Wrist phones are as common as wrist watches. Technology allows people to commute to work only two to three days per week. At home, HDTV has replaced analog TV. Videophones have been in homes for a few years and there is an emerging market for wrist videophones. Many people you know use the wrist videophones that *your company* makes as shown in Fig. A.1. It has a 1 in. diagonal black-and-white display, only 15 minutes of talk time, but 10 hours of standby time. The refresh rate of 2 frames/second is adequate given the relatively small 1 in. display. This is all possible because the capability of microprocessors accelerated in the late 1990s and early 2000s. Now every new electronics

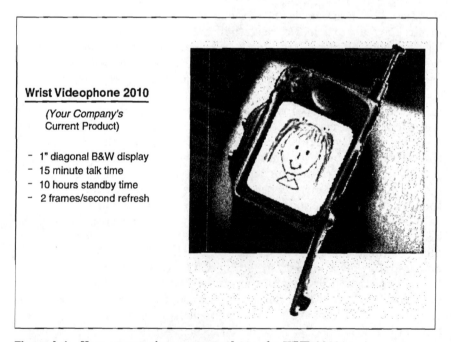

**Figure A.1** *Your company's* current product—the WVP 2010.

# Create a Business with a Strategy

device has a high-speed microprocessor. The personal computer technology leader has new processors that run at 5000 MHz.

## Business Situation

*Your company* is in the rapidly growing wrist phone market. *Your company's* sales of wrist videophones to government agencies have been strong in the last two years and now people in the insurance industry use them. Articles in trade magazines tout that the consumer market will break wide open very soon. *Your company's* President says, "We must capture 30% of these new markets within three years while maintaining high profitability." To do so, *your company* will offer products with better displays than those now offered by Japanese and European consumer products companies and be price-competitive. There are numerous competitive, technology, and cost challenges facing *your company*.

## Competition

Recent sales of wrist videophones far exceeded *your company's* original projections. They are expected to double this year and next. Now that people are very comfortable with cellular phones and use videophones at home, the market for wrist videophones is taking off. Other established companies from watch companies to conglomerates want to get in on the action. There are at least three aggressive startups that will soon offer superior products. *Your company* wants to get established in all market segments before it's too late.

## Technology Challenges

Fig. A.2 shows the subsystems that comprise a typical wrist videophone, and the components that make up those subsystems. Suppliers are touting key components such as high-resolution displays, single-chip phones, miniature speakers, voice-activated "keypads," custom ASICs, etc., suited to this market. The Director of Research believes *your company* has the technology today to offer a superior product that will allow you to maintain market leadership. He proposes to use a new color display that is as vivid as HDTV and which he believes is the next thing consumers will want.

## Cost Challenges

The Marketing VP says, "We need to price our new product 10% below the competition if we are to gain market share." She expects others to cut prices several times in the next three years. So far, *your company* has maintained over 50% margins for the current platform, but she says, "That won't last long. Margins will erode by 5% each year." Since *your company* doesn't continue to make products that drop below 40% gross margin, you need to develop a new platform now. Fig. A.3, the income

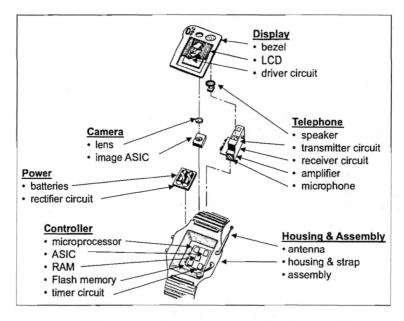

**Figure A.2**  Subsystems and components in a typical wrist videophone.

|  | 2009 Results | |
|---|---|---|
|  | ($M) | (%) |
| **Revenues** | $41.970 | |
| **COGS** | | |
| Cost of Goods Sold | $17.872 | |
| **Gross Margin** | $24.098 | 57% |
| **Direct Expenses** | | |
| Marketing & Sales | $4.700 | |
| Distribution | $2.850 | |
| Research & Development | $6.345 | |
| General & Administrative | $2.900 | |
| **Allocated Expenses** | | |
| Total Allocated | $2.110 | |
| Provision for Uncollectables | $0.825 | |
| **Other Income** | | |
| Total Other Income | $0.200 | |
| **Operating Income** | $4.568 | 11% |
| (% Return on Sales) | | |

**Figure A.3**  *Your company's* 2009 income statement.

# Create a Business with a Strategy

statement published in the 2009 annual report, shows that overall gross margin for *your company* was 57%.

## Your Objective

You have one year to develop the next-generation product platform that *your company* will produce for the following three years. Maximize *your company's* financial performance for the four-year period (2000–2003) by introducing the right product platform and meeting the target cost.

## The Situation

This is an exercise where you achieve a net financial return based on the results of a market simulation. There are five products in the simulator for each year:

- *Your company's* current product (which is in decline).
- *Your company's* new product (which will gradually supersede your current product).
- A product from each of three confrontational competitors who continually adapt to the market needs.

The simulation covers four years:

- The current year (2010) during which *your company* develops the new product.
- Three future years (2011–2013) when your new product is available in the market.

The five products will be sold over the four-year period into three market segments:

- Government
- Business
- Consumer.

*Your company's* market share in each year will be the total of both the current and the new product. The situation is illustrated in Fig. A.4.

## ACTIVITY: CREATE A COMPANY WITH A STRATEGY

It is time for you to begin your work in the Exercise. The first thing you must decide is, "What is my company's goal, and what is its strategy in this marketplace?" We have made the first part easy for you—the company exists, it's in the wrist videophone business, and it already has a successful product. All you have to do is invent a name for the company.

The second part—the strategy—is harder. Basically, we want you to consider the addressable markets, what market segments (if any) you want to focus on, think about how you compare to the competition, and

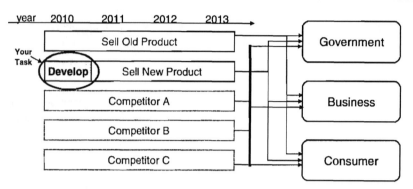

**Figure A.4** The five products over the four years, being sold into three market segments.

have some idea about how you will apply your strengths and weaknesses. Right now you don't have enough information to do this.

### Obtaining Data and Information

So how do you get the information? Simply put, *your company* buys it from a variety of sources (consultants, etc.). Each module requires that you select the information that you think you need from a menu, and record its cost. These costs will be considered your "development expenses" as you progress toward developing and bringing your product to market. Through the Exercise, you can spend as little as $540,000 or as much as $3,590,000. (This reflects the fact that in the business world, you have to pay for information—either consultants provide it, or you get it from subscription services, or your staff has to go out and get it.)

As is often the case in the real world, you may not have all the information you want until after you have paid for it. And as is also the case in the real world, you are sometimes disappointed with the data after you pay for it. The authors have not attempted to defraud you, but some of the data may not apply to the strategy you have chosen. Also, if you are in a later module and feel you now need an item of information that you didn't purchase from a previous module, it's OK to go back and obtain it—just be sure to mark the box and record the expense.

If you are in a classroom setting, it is more effective if the instructor has copies of the information items and "sells" them to the teams as they request them. To stimulate market pressures that require rapid decision-making, we have found it helpful to let the teams know that all prices are doubled after the first 10 minutes.

### Honor System

The information pertaining to each module can be found in the pages at the end of each module. If you decide that you want an item, turn to the

# Create a Business with a Strategy

appropriate page and *check the box* in the corner of the page. This says that you used the information. We assume that your integrity will compel you to check off a page if you look at the information on it. Remember, if you look at any of the information at the end of the chapter, you have bought it! The cost of using this information will be included in the financial statements.

### Information Menu for Module A

To do this part of the Exercise, we offer nine items for your consideration. The menu of items begins on page 145, just before the item pages themselves. Prices are as indicated. You *must buy at least three items* from this list.

### Your Task for Module A

(Time limit: 45 minutes.) Purchase the information that you think you need (three items minimum), invent a name for *your company*, decide on a strategy, and fill in Table A.1. *Note*: You can change your strategy at any time during the Exercise, as you get additional information.

### After Completing Module A

If you are in a classroom setting, the teams should share their strategy and rationale with each other. The discussion can be quite valuable. Either as an individual reader or as a member of a class, depending on what information you obtained, here are some of the things you should have learned at the end of Module A:

- Your business scenario.
- *Your company's* product and name.
- Your business strategy.
- Last year's financial results.
- Market forecasts—sizes, shares.
- Competitive strengths and weaknesses.
- Competitors' costs.

> You have completed Module A of the Exercise.
> Please return to Chapter 2, page 15.

### REFERENCE

1. Laseter, Timothy M., Ramachandran, C.V., and Voigt, Keith H., "Setting Supplier Cost Targets: Getting Beyond the Basics," *Strategy and Business*, issue 6, 1st qtr. 1997, pp. 4–10.

**TABLE A.1** Output for Module A—Create a Company with a Strategy

---

Your Company's Name: _____

Your Strategy:
    Intended sales of the <u>new</u> product into each segment:
        Gov't.: ____ %    Bus.: ____ %    Cnsm.: ____ %    Total=100%

    How this will be done (3-4 sentences):




    Information that supports this strategy (2-4 sentences):




**Information Purchased In Module A** (fill in the cost for the ones that you purchased)

| Item Letter (A, B, C, ....) | Cost ($) |
|---|---|
| A | |
| B | |
| C | |
| D | |
| E | |
| F | |
| G | |
| H | |
| I | |
| Total = | |

---

## INFORMATION MENU FOR MODULE A

We offer the following items for your consideration. Prices are as indicated. You *must buy at least three items* from this list. You will pay between $150,000 and $650,000 for the information that you select.

Market Analysis Information

    A.   DataSearch market research forecast. ($50,000)

# Create a Business with a Strategy

   B. Independent business consultant's market forecast. ($50,000)
   C. Market share by competitor. ($50,000)
   D. Market share in each segment for each company. Government segment is *your company* = 55%, Company A = 25%, Company B = 15%, Company C = 5%.) ($100,000)
   E. Percentage of sales by segment for each company. *Your company* is Government = 60%, Business = 30%, Consumer = 10%.) ($100,000)

Competitive Analysis Information

   F. Strengths, weakness, and strategy for each company. ($50,000)
   G. Reverse engineering of competitors' costs. ($150,000)
   H. Benchmarking report on costs and practices for similar products. ($200,000—includes G)
   I. Consultant's view about the competitors. ($250,000—includes G and H)

**Item A  DataSearch Market Research Forecast**

Shifting market reported by responding companies:

- Government is the largest market segment now.
- Business will have the largest short-term growth.
- Consumer has the greatest market potential.

Module A
Info. Item A

Yes, I looked at this item

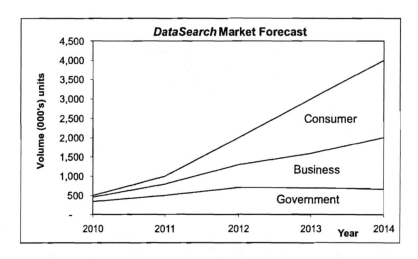

| Segment | DataSearch Market Forecast  # units (000's) |       |       |       |       |       |
|---|---|---|---|---|---|---|
|  | 2009 | 2010 | 2011 | 2012 | 2013 | 2014 |
| Government | 115 | 350 | 500 | 700 | 700 | 650 |
| Business | 65 | 100 | 300 | 600 | 900 | 1,350 |
| Consumer | 20 | 50 | 200 | 700 | 1,400 | 2,000 |

# Create a Business with a Strategy

**Item B** Independent Business Consultant's Market Forecast

Module A
Info. Item B

Dallas Consulting Group says:
"Shifting market—big shift comes later; three-year view is more realistic."

Yes, I looked at this item

- Government—never declines, but grows at slower rate.
- Business—lower early, but more rapid growth comes in 2012.
- Consumer—relatively small market, but bigger when it takes off.

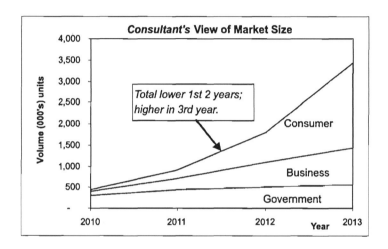

| Segment | Consultant's View # units (000's) | | | | |
|---|---|---|---|---|---|
| | 2010 | 2011 | 2012 | 2013 | 2014 |
| Government | 300 | 450 | 500 | 575 | 600 |
| Business | 100 | 250 | 600 | 850 | 1,350 |
| Consumer | 50 | 200 | 700 | 2,000 | 3,500 |
| Total Volume | 450 | 900 | 1,800 | 3,425 | 5,450 |

**Item C   Market Share by Competitor**

DataSearch estimates:

*Your Company*:
    Largest share; strong in government sales.
Competitors
    Strong in other segments.
    Two large competitors, many small ones.

Module A
Info. Item C

Yes, I looked
at this item

### % Share Total (2010)

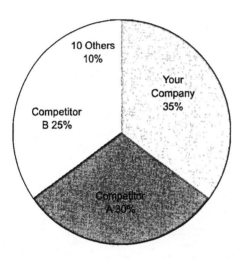

# Create a Business with a Strategy

**Item D** Market Share in Each Segment for Each Company

Module A
Info. Item D

*Your Company*:
  Largest share; strong in government sales.
Competitors:
  Strong in other segments.
  Two large competitors, many small ones.

**Yes, I looked at this item**

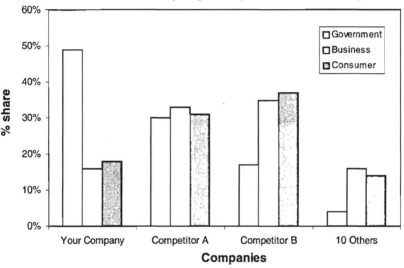
Percent Share by Segment (based on 2010 sales)

Item E  Percentage of Sales by Segment for Each Company

Module A
Info. Item E

Yes, I looked at this item

*Your Company*:
    Largest share; strong in government sales.
Competitors:
    Strong in other segments.
    Two large competitors, many small ones.

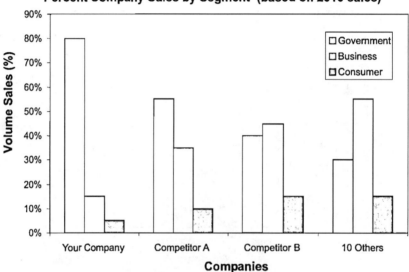

## Create a Business with a Strategy

**Item F  Strengths, Weaknesses, and Strategy for Each Company**

Module A
Info. Item F

Some possible strategies:

Yes, I looked at this item

- Low risk—focus on government (the one *your company* knows).
- Medium risk—focus on business (similar needs and growing).
- High risk—focus on consumer (greatest long-term potential).

|  | Strategy | Strength | Weakness |
|---|---|---|---|
| *Your Company* (35% share) | Become confrontational. Leverage the Gov't market. Gain leadership in other areas using advanced displays. | Know Government Segment. Voice technology. Display technology. | Lack of focus. |
| Company A (30% share) | Product Leadership: Maintain leadership with entries at all price points. | Know business segments. Re-usable platforms. Radio technology. | Patent position. |
| Company B (25% share) | Operational Experience: Be lowest-cost provider. Grow into Business market. | Dominate real-estate applications. Efficient operations keeps costs down. | Slow to adopt technology. |
| 10 Others (10% share) | Grow through acquisitions. Follow market shifts. | Customer experience. Large watch companies. | Fragmentation. |

## Item G  Reverse Engineering of Competitor's Costs

Module A
Info. Item G

Four products from top competitors compared:

Yes, I looked at this item

| Competitor | Product Name | Subsystem | Description | COGS | Price |
|---|---|---|---|---|---|
| A | BestWatch | Display | 1.5" color | $ 30 | |
| | | Controller | chip-on-glass | $ 25 | |
| | | Telephone | 2-chip solution | $ 30 | |
| | | Power | LI-Ion batteries | $ 25 | |
| | | Other | chrome | $ 190 | |
| | | | | $ 300 | $ 600 |
| A | BetterWatch | Display | 1.5" B&W | $ 18 | |
| | | Controller | chip-on-glass | $ 25 | |
| | | Telephone | 2-chip solution | $ 30 | |
| | | Power | NiMH btteries | $ 20 | |
| | | Other | ---- | $ 182 | |
| | | | | $ 275 | $ 575 |
| B | Accel-2 | Display | 1.0" B&W | $ 15 | |
| | | Controller | multi-chip module | $ 75 | |
| | | Telephone | ---- | $ - | |
| | | Power | NiCd batteries | $ 15 | |
| | | Other | ---- | $ 150 | |
| | | | | $ 255 | $ 550 |
| C | Timely-1 | Display | 1.0" B&W | $ 14 | |
| | | Controller | multi-chip-module | $ 85 | |
| | | Telephone | in-controller module | $ - | |
| | | Power | NiMH batteries | $ 15 | |
| | | Other | ---- | $ 211 | |
| | | | | $ 325 | $ 540 |

# Create a Business with a Strategy

**Item H  Benchmarking Report on Costs and Practices for Similar Products**

Module A
Info. Item H

Four products from top competitors compared:

**Yes, I looked at this item**

| Competitor | Product Name | Subsystem | Description | COGS | Price | Reverse-Engineering Findings |
|---|---|---|---|---|---|---|
| A | BestWatch | Display | 1.5" color | $ 30 | | Display & Controller could be combined. |
| | | Controller | chip-on-glass | $ 25 | | |
| | | Telephone | 2-chip solution | $ 30 | | — |
| | | Power | LI-ion batteries | $ 25 | | Batteries in band. |
| | | Other | chrome | $ 190 | | — |
| | | | | $ 300 | $ 600 | |
| A | BetterWatch | Display | 1.5" B&W | $ 18 | | Display & Controller could be combined. |
| | | Controller | chip-on-glass | $ 25 | | |
| | | Telephone | 2-chip solution | $ 30 | | — |
| | | Power | NiMH btteries | $ 20 | | — |
| | | Other | — | $ 182 | | — |
| | | | | $ 275 | $ 575 | |
| B | Accel-2 | Display | 1.0" B&W | $ 15 | | — |
| | | Controller | multi-chip module | $ 75 | | Low-cost MCM. |
| | | Telephone | — | $ - | | — |
| | | Power | NiCd batteries | $ 15 | | — |
| | | Other | — | $ 150 | | Antenna in band. |
| | | | | $ 255 | $ 550 | |
| C | Timely-1 | Display | 1.0" B&W | $ 14 | | — |
| | | Controller | multi-chip-module | $ 85 | | Partner makes MCM. |
| | | Telephone | in-controller module | $ - | | — |
| | | Power | NiMH batteries | $ 15 | | — |
| | | Other | — | $ 211 | | — |
| | | | | $ 325 | $ 540 | |

Module A
Info. Item I

# Item I  Consultant's View About the Competitors

Strongest competitors will:
- Combine functions into fewer parts.
- Develop supplier partnerships.
- Be able to compete with lower margins.

☐ Yes, I looked at this item

| Competitor | Product Name | Subsystem | Description | Process | COGS | Price | Consultant's View |
|---|---|---|---|---|---|---|---|
| A | BestWatch | Display | 1.5" color chip-on-glass | Display & Controller could be combined. | $ 30 | | Will combine display & controller. |
| | | Controller | | | $ 25 | | |
| | | Telephone | 2-chip solution | — | $ 30 | | Will develop 1-chip telephone. |
| | | Power | LI-Ion batteries | Batteries in band. | $ 25 | | Add'l. talk time from batteries in band desirable. |
| | | Other | chrome | | $ 190 | | — |
| | | | | | $ 300 | $ 600 | |
| A | BetterWatch | Display | 1.5" B&W chip-on-glass | Display & Controller could be combined. | $ 18 | | Will combine display & controller. |
| | | Controller | | | $ 25 | | |
| | | Telephone | 2-chip solution | — | $ 30 | | Will develop 1-chip telephone. |
| | | Power | NiMH btteries | — | $ 20 | | — |
| | | Other | — | — | $ 182 | | — |
| | | | | | $ 275 | $ 575 | |
| B | Accel-2 | Display | 1.0" B&W multi-chip module | — | $ 15 | | — |
| | | Controller | | Low-cost MCM. | $ 75 | | Combine teleph. & contr. saves cost. |
| | | Telephone | — | — | $ — | | — |
| | | Power | NiCd batteries | — | $ 15 | | — |
| | | Other | — | Antenna in band. | $ 150 | | — |
| | | | | | $ 255 | $ 550 | |
| C | Timely-1 | Display | 1.0" B&W multi-chip-module | Partner makes MCM. | $ 14 | | — |
| | | Controller | in-controller module | — | $ 85 | | — |
| | | Telephone | — | — | $ — | | Will combine transmitter/receiver. |
| | | Power | NiMH batteries | — | $ 15 | | — |
| | | Other | — | — | $ 211 | | Partners with only 2 suppliers. |
| | | | | | $ 325 | $ 540 | |

# Module B

## Quantify Customers' Needs

This part of the Exercise could also have been titled "Determine What the Customers Want." You need to determine the relative importance—to the customers—of a number of attributes or characteristics for a wrist videophone. Upon completion of this module, you should be able to:

- Summarize relevant product feature information with a market-feature table.
- Assign relative importance to each function or attribute, based on buyer utilities.
- Relate the functions of *your company's* product to the context of *your company's* strategy.
- Select functions, features, and characteristics that are likely to give *your company's* new product the maximum appeal to your selected market segments.

### ATTRIBUTES OF A WRIST VIDEOPHONE

From its experience in the industry *your company* knows that, from the customers' viewpoints, a wrist videophone has eight attributes that need to be considered. They are:

1. *Color*—the color of the band and the bezel surrounding the display, as seen when worn on the wrist.
2. *Display*—includes the size of the video display, and whether it is black-and-white or color.
3. *Speed*—the speed at which the display is refreshed (i.e., the display's refresh rate in frames per second).

4. *Stored numbers*—how many phone numbers can be stored in memory, for speed-dialing purposes.
5. *Talk time*—the amount of time that the wrist videophone can be used before the batteries go flat.
6. *Message service*—the ability to receive and play short voice messages.
7. *Delivery*—the time between placing an order for, and receiving, a wrist videophone.
8. *Price*—the price that the user pays *your company* for a wrist videophone.

## YOUR OBJECTIVE

Determine the relative importance to the customers of the *first seven* of these attributes. (Omit price—it is a major buying-decision factor, but you will determine price independently of the features' relative importance.) Use the information to highlight the top three in importance. You can use one (or a combination) of the following:

- Make estimates from marketplace, competitor, and consultant information.
- Use buyer utilities from a Conjoint Analysis.
- Use range-of-difference in buyer utilities (similar to electric pencil sharpener example in Chapter 2).

Also, select the set of features that you believe will most likely lead to high sales in the market segments that you are focusing on.

## INFORMATION MENU FOR MODULE B

To do this part of the Exercise, we offer eight new items for your consideration. The menu of items can be found on page 159, just before the item pages themselves. Prices are as indicated. You *must buy at least the first item* from this list.

*Note*: If the price of the information seems high, just consider it in the context of the size of your business ($42 million in 2009). Also remember that it does cost money to evaluate a supplier, or to collect market data, or to support a marketing organization that collects market data. The cost for each piece of information simulates these real business costs and incorporates them into the product cycle.

## YOUR TASK FOR MODULE B

(Time limit: 60 minutes.) Purchase the information you think you need (first item minimum), determine the relative importance of all attributes except price, and fill in Table B.1. Then use Table B.2 to select the set of features for *your company's* new product so that it will have the maximum

# Quantify Customers' Needs

**TABLE B.1** Output for Module B (Part 1)—Quantify Customers' Needs

---

Your Company's Name: _____

**Relative Importance of Key Features or Attributes** (for the New Product in its Marketplace):
(NOTE: Price is excluded.)

| Functional Need | Relative Importance (%) |
|---|---|
| Color | |
| Display | |
| Speed | |
| Stored Numbers | |
| Talk Time | |
| Message Service | |
| Delivery | |
| Total = | 100% |

**The Top 3 Attributes:**

1. _____
2. _____
3. _____

**Information Purchased In Module B** (fill in the cost for the ones that you purchased):

| Item Letter (A, B, C, ....) | Cost ($) | |
|---|---|---|
| A | $50K | (REQUIRED) |
| B | | |
| C | | |
| D | | |
| E | | |
| F | | |
| G | | |
| Total = | | |

---

appeal in the market segments that are consistent with your strategy. (*Note*: Consider this a preliminary selection of features; you may choose to change one or two of them when you get to Module E.)

## AFTER COMPLETING MODULE B

If you are in a classroom setting, the teams should share their lists of relative importance of attributes with each other. The discussion can be

**TABLE B.2** Output for Module B (Part 2)—Select Product Features

---

Your Company's Name: _____

**Features to be Provided In Your Company's New Product:**
(NOTE: These may not be your final selections. You could decide to change one or two of them later, when you are in Module E.)

| Attribute, and Level | Circle <u>one</u> number in each group |
|---|:---:|
| **Color:** | |
| Chrome | 1 |
| Black (standard) | 2 |
| Custom: brown, blue, other | 3 |
| Gold | 4 |
| **Display:** | |
| 1.5 inch diagonal, black & white | 1 |
| 1.5 inch diagonal, color | 2 |
| 2.0 inch diagonal, color | 3 |
| **Video Speed:** | |
| industry standard (2 frames/second) | 1 |
| standard videophone (12 frames/second) | 2 |
| computer movie quality (24 frames/second) | 3 |
| **Stored Numbers:** | |
| 4 numbers | 1 |
| 16 numbers | 2 |
| 32 numbers | 3 |
| **Talk Time:** | |
| 15-minute talk time, 10-hr. standby | 1 |
| 30-minute talk time, 16-hr. standby | 2 |
| 60-minute talk time, 24-hr. standby | 3 |
| **Messaging Service:** | |
| None | 1 |
| stores 2 minutes of voice messages | 2 |
| stores 5 minutes of voice messages | 3 |
| **Delivery:** | |
| within 24 hours | 1 |
| within 72 hours | 2 |

---

quite valuable. Either as an individual reader or as a member of a class, depending on what information you learned, here are some of the things you should have learned at the end of Module B:

In Module A

- Your business scenario.
- *Your company's* product and name.
- Your business strategy.
- Last year's financial results.

## Quantify Customers' Needs

- Market forecasts—sizes, shares.
- Competitive strengths and weaknesses.
- Competitors' costs.

In Module B

- Customer preferences.
- Market-feature table.
- Conjoint Analysis: buyer utilities.
- Product attributes: importance-value table.
- Features that will be in your product.

> You have completed Module B of the Exercise.
> Please return to Chapter 3, page 39.

## INFORMATION MENU FOR MODULE B

We offer the following items for your consideration. Prices are as indicated. You *must buy at least the first item* from this list. You will pay between $50,000 and $450,000 for the information that you select.

A. Customer preferences worksheet (all attributes and levels). REQUIRED. ($50,000)
B. Market-feature table for government customers. ($50,000)
C. Market-feature table for business customers. ($50,000)
D. Market-feature table for consumer customers. ($50,000)
E. Market-feature table for all customers (includes B, C, and D). ($100,000)
F. Consultant's view on attributes and levels used in marketing study. ($50,000)
G. Complete Conjoint Analysis marketing study. ($250,000)

## Item A  Customer Preferences Worksheet
(All Attributes and Levels)

Module B
Info. Item A

Fill in buyer utilities relevant to your product strategy.
Decide how you want to weight these attributes in your product.

Yes, I looked at this item

| Wrist Videophone | Buyer Utilities | Max-Min | Relative Weight | Rel. Wt. (w/o Price) |
|---|---|---|---|---|
| **Color:** | | | | |
| Chrome ($5 discount) | | | | |
| Black (standard) | | | | |
| Custom color exterior: brown, blue or other for $25 | | | | |
| Gold case for $35 | | | | |
| **Display:** | | | | |
| 1.5 inch diagonal black & white | | | | |
| 1.5 inch color for $25 | | | | |
| 2.0 inch color for $50 | | | | |
| **Video Speed:** | | | | |
| Industry standard: 2 frames/sec. ($25 discount) | | | | |
| Standard videophone: 12 frames/sec. | | | | |
| Computer movie quality: 24 frames/sec. For $50 | | | | |
| **Stored Numbers:** | | | | |
| 4 | | | | |
| 16 for $10 | | | | |
| 32 for $25 | | | | |
| **Talk Time:** | | | | |
| 10 hr. standby, 15 minute talk time | | | | |
| 24 hr. standby, 60 minute talk time for $75 | | | | |
| 17 hr. standby, 30 minute talk time for $40 | | | | |
| **Messaging Service:** | | | | |
| None | | | | |
| Store 2 minutes of messages for $25 | | | | |
| Store 5 minutes of messages for $40 | | | | |
| **Delivery:** | | | | |
| 24 hours for shipping premium | | | | |
| 72 hours | | | | |
| **Selling Price:** | | | | |
| $150 | | | | |
| $250 | | | | |
| $350 | | | | |
| $450 | | | | |
| $550 | | | | |

## Item B   Market-Feature Table for Government Customers

Module B
Info. Item B

Qualitative measures of features preferred by *government* customers:

Yes, I looked at this item

|  | Government |
|---|---|
| Premium | store 32 #s<br>48 hr. standby<br>5 min. messages<br>password |
| Step-Up | 2.0" color display<br>24 frames/sec.<br>24 hr. standby<br>60 min talk time<br>2 min. messages |
| Basic | black<br>1.5" color display<br>12 frames/sec.<br>16 hr. standby<br>30 min. talk time<br>store 16 #s<br>no messaging<br>72 hr. delivery |

### Item C  Market-Feature Table for Business Customers

Module B
Info. Item C

Yes, I looked at this item

Qualitative measures of features preferred by *business* customers:

|  | Insurance/Sales | Orther Business |
|---|---|---|
| **Premium** | gold case<br>48 hr. standby<br>password | 2.0" color display<br>12 frames/sec.<br>Store 32 #s<br>24 hr. delivery |
| **Step-Up** | custom color<br>1.5" color display<br>12 frames/sec.<br>24 hr. standby<br>60 min talk time<br>store 32#s<br>5 min. messages<br>24 hr. delivery | custom color<br>16 hrs. standby<br>60 min. talk time<br>store 16 #s<br>2 min. messages |
| **Basic** | chrome<br>1.5" B&W display<br>2 frames/sec.<br>16 hr. standby<br>60 min. talk time<br>store 16 #s<br>2 min. messages<br>72 hr. delivery | chrome<br>1.5" color display<br>2 frames/sec.<br>10 hr. standby<br>15 min. talk time<br>store 4 #s<br>No messaging<br>72 hr. delivery |

## Item D  Market-Feature Table for Consumer Customers

Module B
Info. Item D

Qualitative measures of features preferred by *consumer* customers:

Yes, I looked at this item

| | Consumer |
|---|---|
| **Premium** | gold case<br>2.0" color display<br>16 hrs. standby<br>30 min. talk time<br>2 min. messages |
| **Step-Up** | custom color<br>1.5" color display<br>12 frames/sec.<br>store 16 #s |
| **Basic** | chrome<br>1.5" B&W display<br>2 frames/sec.<br>10 hr. standby<br>15 min. talk time<br>store 4 #s<br>No messaging<br>72 hr. delivery |

## Item E  Market-Feature Table for All Customers

Module B
Info. Item E

Qualitative measures of features preferred by *all* customers:

Yes, I looked at this item

|  | Wrist Videophone Market Segments ||||
|  | Government | Insurance/Sales | Other Business | Consumer |
|---|---|---|---|---|
| **Premium** | store 32 #s<br>48 hr. standby<br>5 min. messages<br>password | gold case<br>48 hr. standby<br>password | 2.0" color display<br>12 frames/sec.<br>Store 32 #s<br>24 hr. delivery | gold case<br>2.0" color display<br>16 hr. standby<br>30 min. talk time<br>2 min. messages |
| **Step-Up** | 2.0" color display<br>24 frames/sec.<br>24 hr. standby<br>60 min. talk time<br>2 min. messages | custom color<br>1.5" color display<br>12 frames/sec.<br>24 hr. standby<br>60 min. talk time<br>store 32#s<br>5 min. messages<br>24 hr. delivery | custom color<br>16 hrs. standby<br>60 min. talk time<br>store 16 #s<br>2 min. messages | custom color<br>1.5" color display<br>12 frames/sec.<br>store 16 #s |
| **Basic** | black<br>1.5" color display<br>12 frames/sec.<br>16 hr. standby<br>30 min. talk time<br>store 16 #s<br>no messaging<br>72 hr. delivery | chrome<br>1.5" B&W display<br>2 frames/sec.<br>16 hr. standby<br>60 min. talk time<br>store 16 #s<br>2 min. messages<br>72 hr. delivery | chrome<br>1.5" color display<br>2 frames/sec.<br>10 hr. standby<br>15 min. talk time<br>store 4 #s<br>No messaging<br>72 hr. delivery | chrome<br>1.5" B&W display<br>2 frames/sec.<br>10 hr. standby<br>15 min. talk time<br>store 4 #s<br>No messaging<br>72 hr. delivery |

**Item F  Consultant's View on Attributes and Levels Used in Study**

Module B
Info. Item F

Yes, I looked at this item

The ideal *Government* product is:

- Black.
- Color display.
- Movie quality.
- Stores 16 numbers.
- Has 30-minute talk time.

The ideal *Business* product is:

- Chrome.
- Color display.
- Video speed.
- Stores 16 numbers.
- Has 30-minute talk time.
- Affordable.

The ideal *Consumer* product is:

- Any color.
- Inexpensive.

**Item G1  Complete Conjoint Analysis
Marketing Study (Tabular)**

Module B
Info. Item G1

Yes, I looked
at this item

| Wrist Videophone Buyer Utilities | Govt. Avg. | Bus. Avg. | Cons. Avg. | Overall Avg. |
|---|---|---|---|---|
| **Color:** | | | | |
| Chrome ($5 discount) | 8 | 45 | 38 | 30 |
| Black (standard) | 71 | 6 | 9 | 29 |
| Custom color exterior: brown, blue or other for $25 | 12 | 20 | 40 | 24 |
| Gold case for $35 | 1 | 4 | 14 | 6 |
| **Display:** | | | | |
| 1.5 inch diagonal black & white | 9 | 18 | 39 | 22 |
| 1.5 inch color for $25 | 69 | 60 | 22 | 50 |
| 2.0 inch color for $50 | 42 | 12 | 4 | 19 |
| **Video Speed:** | | | | |
| Industry standard: 2 frames/sec. ($25 discount) | 0 | 43 | 55 | 33 |
| Standard video phone: 12 frames/sec. | 42 | 51 | 25 | 39 |
| Computer movie quality: 24 frames/sec. For $50 | 46 | 1 | 0 | 16 |
| **Stored Numbers:** | | | | |
| 4 | 0 | 24 | 27 | 17 |
| 16 for $10 | 45 | 44 | 14 | 34 |
| 32 for $25 | 15 | 27 | 0 | 14 |
| **Talk Time:** | | | | |
| 10 hr. standby, 15 minute talk time | 0 | 24 | 54 | 26 |
| 24 hr. standby, 60 minute talk time for $75 | 70 | 43 | 34 | 49 |
| 16 hr. standby, 30 minute talk time for $40 | 65 | 22 | 2 | 30 |
| **Messaging Service:** | | | | |
| None | 33 | 21 | 45 | 33 |
| Store 2 minutes of messages for $25 | 17 | 19 | 15 | 17 |
| Store 5 minutes of messages for $40 | 15 | 15 | 0 | 10 |
| **Delivery:** | | | | |
| 24 hours for shipping premium | 0 | 7 | 0 | 2 |
| 72 hours | 23 | 15 | 13 | 17 |
| **Selling Price:** | | | | |
| $150 | 65 | 106 | 181 | 117 |
| $250 | 72 | 88 | 123 | 94 |
| $350 | 48 | 52 | 40 | 47 |
| $450 | 32 | 28 | 7 | 22 |
| $550 | 1 | 4 | 0 | 2 |

# Quantify Customers' Needs

**Item G2** Complete Conjoint Analysis
Marketing Study (Graphical)

Module B
Info. Item G2

Yes, I looked at this item

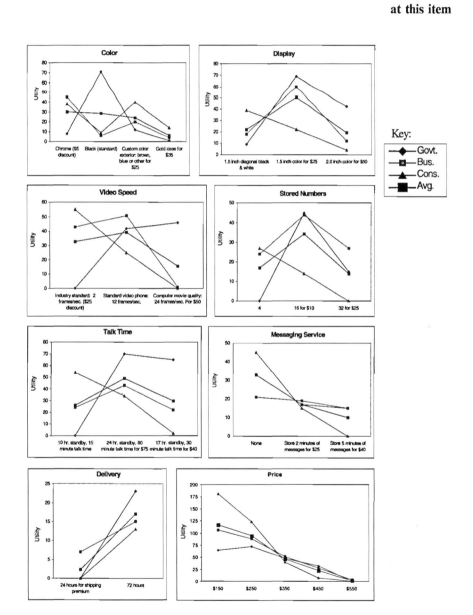

# Module C

## Determine Target Price and Cost (Product Level)

This part of the Exercise will apply the principles covered in Chapter 3. We will focus on how to "Determine Future Prices and Costs with an Experience Curve." You need to estimate the prices your customers will pay in 2011, 2012, and 2013 for your new wrist videophone. You will be guided by the decisions you made in the previous exercises where you arrived at a strategy: Government, Business, Consumer, or a combination of each. You also selected the set of features in the basic product. But remember that the marketplace controls the price of your product. Your task is to predict what that price will be based on the experience curve. Upon completion of this module, you should be able to:

- Create a price/cost experience curve based on market price data and *your company's* gross margin.
- Determine the target cost for your product for a period of time in the future.

**YOUR OBJECTIVE**

Given historical prices, current prices and costs, and trends in similar products, determine the target cost of your wrist videophone for the next three years. We will give you the price and cost of your current product—and the industry cumulative volume—for the year 2009. *Note*: The cumulative volume includes the year 2009 *plus all prior years*, and amounts to a total of 500,000 units.

- Determine the slope of the price curve for the future years, based on price trends of similar products.

- Find the industry cumulative volume for the year 2010. (You will have to calculate the projected cumulative volumes from information that you may have purchased in Module A. Remember that you can always go back and purchase information from previous modules.)
- Find your prices for the current product in the current year 2010. (You will have to purchase that information here in Module C.)
- Starting with the 2010 data, and using the slope you determined, plot the resulting price experience curve on the log-log chart provided.
- Determine the target prices for 2011, 2012, and 2013. Set the target *gross* margins for 2011, 2012, and 2013. (*Note*: You can't go lower than 40%; see Module A, page 139). You can change the gross margin as the product matures. This will slightly change the cost curve but not the price curve you generated.
- Determine the product-level target costs (COGS) for the years 2011, 2012, and 2013. Remember that this product is being designed in 2010 and although you will not need to set a price for the new product in 2010, you will have used the current product's 2010 price as part of the experience curve.

## INFORMATION MENU FOR MODULE C

To do this part of the Exercise, we offer six new items for your consideration. The menu of items can be found on page 172, just before the item pages themselves. Prices are as indicated. You *must buy at least the first item* from this list.

## YOUR TASK FOR MODULE C

(Time limit: 45 minutes.) Purchase the information that you think you need (first item minimum), plot the price and cost experience curves, and list your new product's planned prices, margins, and costs for the next three years. Put your results in Table C.1.

## AFTER COMPLETING MODULE C

As always, if you are in a classroom setting we strongly encourage the teams to share their planned prices, margins, and costs with each other. Either as an individual reader or as a member of a class, depending on what information you obtained, here are some of the things you should have learned by the end of Module C (including a reprise of your learning

# Determine Target Price and Cost (Product Level)

TABLE C.1 Output for Module C—Determine Target Price and Cost (Product Level)

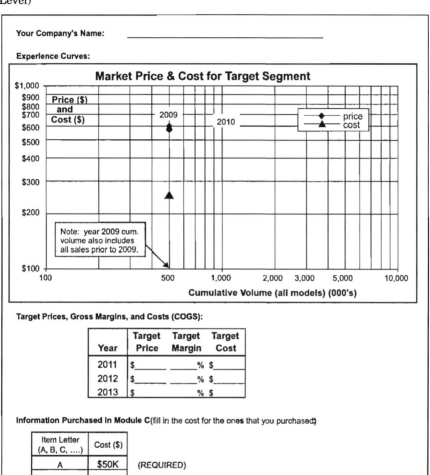

from Modules A and B):

In Module A

- Your business scenario.
- *Your company's* product and name.
- Your business strategy.

- Last year's financial results.
- Market forecasts—sizes, shares.
- Competitive strengths and weaknesses.
- Competitors' costs.

In Module B
- Customer preferences.
- Market-feature table.
- Conjoint Analysis: buyer utilities.
- Product attributes: importance-value table.
- Features that will be in your product.

In Module C
- Historical and current prices for your current product.
- This year's projected financial results.
- Price trends.
- Price experience curve.
- Your planned gross margins.
- Cost experience curve.
- Target prices, costs, and margins for 2011–2013.

> You have completed Module C of the Exercise.
> Please return to Chapter 4, page 55.

## INFORMATION MENU FOR MODULE C

We offer the following items for your consideration. Prices are as indicated. You *must buy at least the first item* from this list. You will pay between $50,000 and $300,000 for the information that you select.

A. *Your company's* historical prices and planned 2010 prices. REQUIRED. ($50,000)
B. *Your company's* first quarter year 2010 income statement. ($50,000)
C. Watch price trends. ($50,000)
D. Cellular phone price trends. ($50,000)
E. Videophone price trends. ($50,000)
F. Consultant's view on price trends for wrist videophones. ($50,000)

## Determine Target Price and Cost (Product Level)     173

**Item A  Baseline Financials (Plan for Costs and Prices of Current Product)**

- *Your company's* historical and planned prices for the current product.
- *Your company's* past and projected costs for the current product.

Module C
Info. Item A

Yes, I looked at this item

| Market Segment | 2009 Price | 2009 Cost | 2010 Price | 2010 Cost |
|---|---|---|---|---|
| Government | $ 600 | $ 254 | $ 525 | $ 220 |
| Business | $ 585 | $ 254 | $ 515 | $ 220 |
| Consumer | $ 575 | $ 254 | $ 505 | $ 220 |

## Item B   Your Company's Current Income Statement

Module C
Info. Item B

*Your company's* Income Statement

- Year 2009 actual.
- Year 2010 actual so far (first quarter).
- Budget for entire year.

Yes, I looked at this item

|  | 2009 | 2010 | |
|---|---|---|---|
|  | Actual | 1Qtr | Plan |
| Revenue ($M) | $ 41.970 | $ 19.400 | $ 81.300 |
| COGS | $ 17.871 | $ 8.200 | $ 32.900 |
|  |  |  |  |
| Gross Margin | $ 24.098 | $ 11.200 | $ 48.400 |
| Gross Margin (%) | 57% | 58% | 60% |
|  |  |  |  |
| **Operating Expenses:** |  |  |  |
| R&D | $ 6.345 | $ 3.000 | $ 9.100 |
| M&S | $ 4.700 | $ 3.000 | $ 8.700 |
| Distribution | $ 2.850 | $ 1.000 | $ 3.800 |
| G&A | $ 2.900 | $ 1.000 | $ 5.200 |
| Total Operating Expenses | $ 16.795 | $ 8.000 | $ 26.800 |
|  |  |  |  |
| **Other Expenses** |  |  |  |
| Other allocated | $ 2.110 | $ 1.000 | $ 5.800 |
| Provision for uncollectiables | $ 0.825 | $ 0.500 | $ 1.500 |
| Total Other Expenses | $ 2.935 | $ 1.500 | $ 7.300 |
|  |  |  |  |
| **NET INCOME** |  |  |  |
| Measured Operating Income | $ 4.568 | $ 1.900 | $ 13.200 |
| ROS (%) | 11% | 10% | 16% |

**Determine Target Price and Cost (Product Level)**

**Item C  Watch Price Trends**

Price experience curve (from 92% to 82% slope over 40 years)

Module C
Info. Item C

Yes, I looked at this item

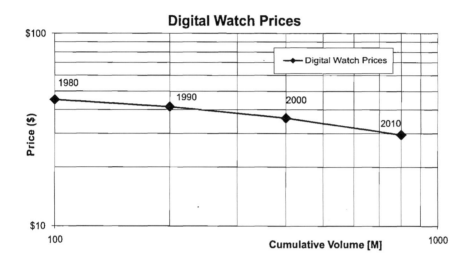

## Item D  Cellular Phone Price Trends

Price experience curve (from 90% to 78% slope over 4 years)

Module C
Info. Item D

Yes, I looked at this item

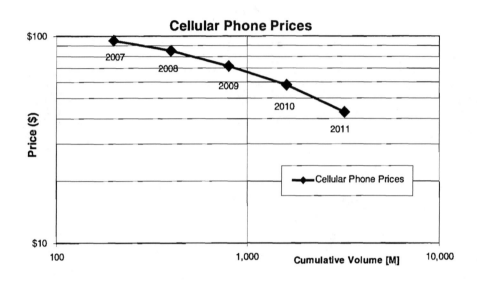

# Determine Target Price and Cost (Product Level) 177

**Item E**   Videophone Price Trends

Price experience curve (from 87% to 75% slope over 4 years)

Module C
Info. Item E

Yes, I looked at this item

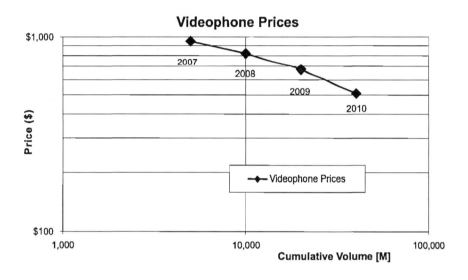

**Item F  Consultant's View on Price Trends for Wrist Videophones**

Module C
Info. Item F

Yes, I looked at this item

Wrist Videophone experience curve slope trends are similar to:

- Digital watch experience curve (~85%) for Government market.
- Cellular phone experience curve (~80%) for Business market.
- Videophone experience curve (~75%) for Consumer market.

# Module D

Determine Cost Targets (Subsystem Level)

In this part of the Exercise you will "Set Subsystem Targets." You need to determine the cost targets of the wrist videophone's subsystems for the years 2011, 2012, and 2013. Upon completion of this module, you should be able to use value analysis to determine the cost targets for the subsystems that comprise a product. For simplicity, we will focus on COGS only.

**YOUR OBJECTIVE**

Determine the cost targets for the subsystems in *your company's* new wrist videophone in the years 2011, 2012, and 2013.

- Subdivide *your company's* new product into its six subsystems (see Fig. A.2). *Note*: The Labor and Load part for putting the wrist videophone together is the "Assembly" part of "Housing and Assembly." All other costs are Material costs.
- Obtain the feature–subsystem Matrix for a wrist videophone. It is in a worksheet that you will purchase (Module D, Item A).
- Use the relative importance of the features (excluding price) that you determined in Module B (Table B.1).
- Use the overall product cost targets for 2011, 2012, and 2013 that you determined in Module C (Table C.1).
- Use value analysis approach and worksheet to calculate subsystem allowable costs in 2011, 2012, and 2013 (same as method shown in Figs. 4.8 and 8.18). *Note*: This involves tedious calculation; you could set up a spreadsheet on your PC and use the "Sumproduct" function.

- Accept these allowable costs as your cost targets for the subsystems in 2011, 2012, and 2013.
- Obtain 2010 costs for the subsystems, or for subsystems and their constituent components. (2010 is the current year, and it is the one for which recent and accurate data exists.)
- Pick an important year in the future (2011, 2012, or 2013) for your strategy and plot the 2010 costs versus that year's cost targets for each of the six subsystems (same as method shown in Figs. 4.9 and 8.19). List the three subsystems that have the most opportunity for improvement (your biggest problem cases).

## INFORMATION MENU FOR MODULE D

To do this part of the Exercise, we offer nine new items for your consideration. The menu of items can be found on page 182, just before the item pages themselves. Prices are as indicated. You *must buy at least the first item* from this list.

## YOUR TASK FOR MODULE D

(Time limit: 60 minutes.) Purchase the information that you think you need (first item minimum), calculate the 2011, 2012, and 2013 subsystem allowable costs, accept them to be the cost targets, and identify the three subsystems that need the biggest cost reduction. Put your results in Table D.1.

## AFTER COMPLETING MODULE D

As always, if you are in a classroom setting we strongly encourage the teams to share their subsystem cost targets with each other. Either as an individual reader or as a member of a class, depending on what information you obtained, here are some of the things you should have learned by the end of Module D:

In Module A
- Your business scenario.
- *Your company's* product and name.
- Your business strategy.
- Last year's financial results.
- Market forecasts—sizes, shares.
- Competitive strengths and weaknesses.
- Competitors' costs.

In Module B
- Customer preferences.
- Market-feature table.
- Conjoint Analysis: buyer utilities.

# Determine Cost Targets (Subsystem Level)

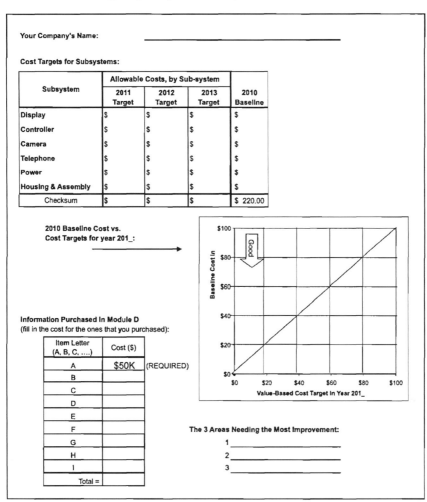

TABLE D.1  Output for Module D—Determine Cost Targets (Subsystem Level)

- Product attributes: importance-value table.
- Features that will be in your product.

In Module C

- Historical and current prices for your current product.
- This year's projected financial results.
- Price trends.
- Price experience curve.
- Your planned gross margins.
- Cost experience curve.
- Target prices and costs (COGS) for 2011–2013.

In Module D
- Function–subsystem matrix.
- Current costs of subsystems and components.
- Cost targets for each subsystem.
- Subsystem cost gaps (2011–2013).

> You have completed Module D of the Exercise.
> Please return to Chapter 5, page 71.

## INFORMATION MENU FOR MODULE D

We offer the following items for your consideration. Prices are as indicated. You *must buy at least the first item* from this list. You will pay between $50,000 and $200,000 for the information that you select.

A. Feature–subsystem matrix and worksheet for value analysis. REQUIRED. ($50,000)
B. *Your company's* 2010 costs for all the subsystems, but not component costs. ($50,000)
C. *Your company's* 2010 costs for the display and each of its components. ($20,000)
D. *Your company's* 2010 costs for the controller and each of its components. ($20,000)
E. *Your company's* 2010 costs for the camera and each of its components. ($20,000)
F. *Your company's* 2010 costs for the telephone and each of its components. ($20,000)
G. *Your company's* 2010 costs for the power and each of its components. ($20,000)
H. *Your company's* 2010 costs for the housing and assembly and each of its components. ($20,000)
I. *Your company's* 2010 costs for all the subsystems, *and* all component costs (includes C, D, E, F, G, and H). ($100,000)

# Determine Cost Targets (Subsystem Level)

**Item A  Feature–Subsystem Matrix and Worksheet for Value Analysis**

Module D
Info. Item A

Yes, I looked at this item

- Put data in the shaded cells:
  The relative importance of the features to your customers.
  The target costs for *your company's* new product in 2011, 2012, and 2013.
- Calculate the value-based cost of each feature in 2011, 2012, and 2013.
- Calculate the resulting cost of each subsystem in *your company's* new product in 2011, 2012, and 2013.
- Put data in the shaded cells:
  The subsystem costs for *your company's* current product in 2010.

(*Note*: If you do the calculations for a single year, the results for the other years will simply scale with the target costs for those years. That is because we are assuming that the relative importance of each feature remains constant over the three-year period. See next page for worksheet.)

## Item A Worksheet

| Subsystem | Customers' Functional Needs ||||||||||||| Allowable Costs, by Subsystem ||||
|---|---|---|---|---|---|---|---|---|---|---|---|---|---|---|---|---|
| | Color | % | Display | % | Video Speed | % | Stored Numbers | % | Talk Time | % | Messages | % | Delivery | % | 2011 Target | 2012 Target | 2013 Target | 2010 Baseline |
| Customer Importance Weighting | % | | % | | % | | % | | % | | % | | % | | Data In | | | Data In |
| Value-Based Costs (2011) | $ | | $ | | $ | | $ | | $ | | $ | | $ | | $ | | | |
| Value-Based Costs (2012) | $ | | $ | | $ | | $ | | $ | | $ | | $ | | | $ | | |
| Value-Based Costs (2013) | $ | | $ | | $ | | $ | | $ | | $ | | $ | | | | $ | |
| Display | 60% | | 65% | | 25% | | — | | 10% | | — | | 10% | | $ | $ | $ | $ |
| Controller | — | | — | | 50% | | 50% | | 30% | | 40% | | 10% | | $ | $ | $ | $ |
| Camera | — | | 15% | | 15% | | — | | — | | — | | 10% | | $ | $ | $ | $ |
| Telephone | — | | — | | 10% | | 20% | | 30% | | 40% | | 10% | | $ | $ | $ | $ |
| Power | — | | 20% | | — | | 30% | | 20% | | 15% | | 10% | | $ | $ | $ | $ |
| Housing & Assembly | 40% | | — | | — | | — | | 10% | | 5% | | 50% | | | | | |
| Checksum | 100% | | 100% | | 100% | | 100% | | 100% | | 100% | | 100% | | $ | $ | $ | $ 220.00 |

**Determine Cost Targets (Subsystem Level)**　　　　　　　　　185

**Item B** Your Company's 2010 Costs for All the Subsystems (But Not Component Costs)

Module D
Info. Item B

Yes, I looked at this item

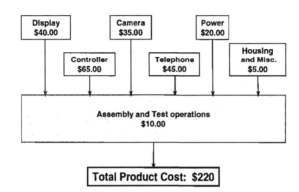

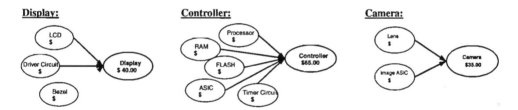

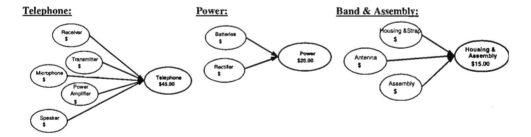

**Module D**

**Item C** Your Company's 2010 Costs for
the Display and Each of Its Components

Module D
Info. Item C

Yes, I looked
at this item

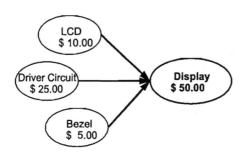

**Item D** Your Company's 2010 Costs for the
Controller and Each of Its Components

Module D
Info. Item D

Yes, I looked
at this item

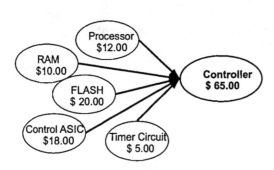

**Determine Cost Targets (Subsystem Level)** 187

**Item E** Your Company's 2010 Costs for the
Camera and Each of Its Components

Module D
Info. Item E

Yes, I looked
at this item

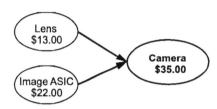

**Item F** Your Company's 2010 Costs for the
Telephone and Each of Its Components

Module D
Info. Item F

Yes, I looked
at this item

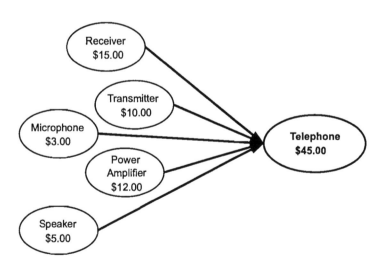

**Item G** Your Company's 2010 Costs for the Power and Each of Its Components

Module D
Info. Item G

Yes, I looked at this item

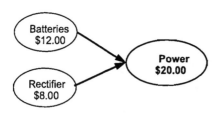

**Item H** Your Company's 2010 Costs for the Housing and Assembly and Each of Its Components

Module D
Info. Item H

Yes, I looked at this item

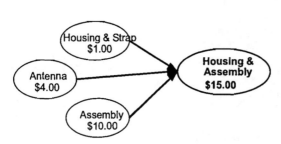

**Determine Cost Targets (Subsystem Level)**

**Item I** Your Company's 2010 Costs for All the Subsystems and All Component Costs

Module D
Info. Item I

Yes, I looked at this item

| Subsystem/Component | Cost in 2010 | | |
|---|---|---|---|
| **Display** | $ 40.00 | | |
|    LCD | | $ | 10.00 |
|    Driver Circuit | | $ | 25.00 |
|    Bezel | | $ | 5.00 |
| **Controller** | $ 65.00 | | |
|    Processor | | $ | 12.00 |
|    RAM | | $ | 10.00 |
|    FLASH | | $ | 20.00 |
|    ASIC | | $ | 18.00 |
|    Timer Circuit | | $ | 5.00 |
| **Camera** | $ 35.00 | | |
|    Lens | | $ | 13.00 |
|    Image ASIC | | $ | 22.00 |
| **Telephone** | $ 45.00 | | |
|    Speaker | | $ | 5.00 |
|    Microphone | | $ | 3.00 |
|    Transmitter | | $ | 10.00 |
|    Receiver | | $ | 15.00 |
|    Power Amplifier | | $ | 12.00 |
| **Power** | $ 20.00 | | |
|    Batteries | | $ | 12.00 |
|    Rectifier Circuit | | $ | 8.00 |
| **Housing & Assembly** | $ 15.00 | | |
|    Housing & Strap | | $ | 1.00 |
|    Antenna | | $ | 4.00 |
|    Assembly | | $ | 10.00 |
| **Total =** | $ 220.00 | $ | 220.00 |

# Module E

Find Paths to Achieve the Targets

In this part of the Exercise you will determine the costs (COGS) of *your company's* new product. In Module D you established cost targets for each of the subsystems in the years 2011, 2012, and 2013. At the end of this module you should be able to:

- subdivide the subsystems into their constituent components;
- find components that will help you achieve the cost targets.

**SCENARIO**

It is difficult in an exercise such as this, either for an individual reader or in a classroom setting, to simulate all the things that might be done to converge on a product design that meets the targets. We cannot expect you to do the detailed engineering design of a wrist videophone, develop new cost-saving technology, work with suppliers to help them reduce the costs of their components, negotiate supplier contracts, or re-engineer the assembly and test processes. So we have arranged this module so that you act as the component engineer and your job is to select the components. Basically, you will have to get quotations for components that are likely to be suitable, and select the set of components that best satisfies your product requirements and meets the cost targets. The quotations come from within you own company ("in house") or four outside suppliers (called "A", "B", "C," and "D").

This module requires the most work, and it has more complex "rules," so we have allowed a much longer time limit. If you are in a classroom setting, we suggest that each team appoints each member to be a "subject-matter expert" (SME) for one of the six subsystems (display, controller, etc.). If there are only five people on a team, we suggest one

191

person can be the SME for both the camera and the housing and assembly. If there are only four, another person can do both the display and the camera.

Naturally, there is a cost associated with component engineering, the quotation process, and working with suppliers, so you will have to purchase each quotation at a price of $20,000 each. Beginning on page 202, you will find a menu of 97 possible quotations. The quotations themselves are numbered; Figure E.1 shows what an example quotation (not one that's in the menu) looks like. You will see that one of the "components" is "assembly and test." This is the labor and load associated with doing the final assembly and testing of the product. There are choices here, because you may do it within your own company ("in house") or have it done on the outside by contract manufacturers.

## REACHING THE TARGET COST

Depending on your targets and other factors, it may be impossible for all subsystems to individually reach their targets. But the important thing is that the overall system reaches its target cost. Therefore some subsystems may have to do much better than their targets and thereby subsidize others.

You may also want to consider the features that you are offering—those that you selected in Table B.2. You may find that you can save considerable costs if you change the features slightly. There is always a risk that the product might be less attractive to the market segment(s) that you are focusing on, but the impact on sales may be small, especially if the feature affected is of low relative importance to the customers. Or you may have selected a feature not needed in the "Basic" product, i.e. one that should have been optional—"Step-Up" or "Premium."

In virtually all cases, you will have to make several iterations or rounds of choices as you converge on a final selection of components. We have provided a worksheet (page 198) that you can use to record your successive component choices and costs. We suggest that you focus on one of the future years (2011, 2012, 2013) and optimize your component selection for that year. When you have finished, record your component choices and record the costs for all years and insure that you have achieved (or nearly achieved) your overall target cost for all years.

On the worksheet, also record and add up the "yearly expenses" associated with each selected component. This simulates that fact that there are ongoing expenses associated with using components—engineering, purchasing, transportation, accounts payable, quality control, reliability, etc. We set these expenses to be the same for all three years (2011, 2012, 2013).

# Find Paths to Achieve the Targets

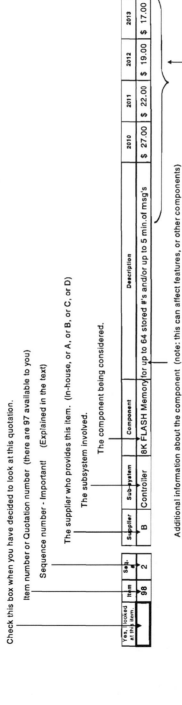

**Figure E.1** Example of a quotation (*Note:* This is not one of the quotations that you can use.)

## SUPPLIER DISCOUNTS

To add one more element of realism, we have added incentives that relate to supplier partnerships. Any company needs to strike an appropriate balance between supplier proliferation and sole-sourcing. With more than one supplier for a component or group of components, it is often possible to stimulate more competitiveness among the suppliers and thereby get lower prices. With fewer suppliers, or even a single supplier, it is more possible to enter into long-term, strategic, and profitable relationships. Both approaches have advantages and disadvantages, but for the purposes of this exercise we have chosen to emphasize the latter approach—fewer suppliers. If you use a small total number of outside suppliers (do not count "in house"), you will get the following discounts:

- If you used a *single* outside supplier for all components, take a *5%* "partnership discount" on the total COGS.
- If you used *two* outside suppliers for all components, take a *2%* "partnership discount" on the total COGS.

Also, you can get a volume discount on the COGS of *one* (and only one) of the subsystems for which all the components come from a *single* supplier, according to the following schedule (pick the one that saves you the most money):

- If a subsystem with a single supplier has *five* equivalent components, take a *20%* "volume discount" on that subsystem's COGS. (This can apply only to the controller or the telephone.)
- If a subsystem with a single supplier has *four* equivalent components, take a *10%* "volume discount" on that subsystem's COGS. (This can apply only to the display or the housing and assembly.)

For example, if you used a single outside supplier for the entire wrist videophone, you can use the 20% discount on the controller or the telephone, *and* then the 5% discount on the entire product.


## INFORMATION MENU FOR MODULE E

To do this part of the Exercise, we offer 97 new items for your consideration. They are the quotations for a variety of components that could be used in your new product. The menu of items begins on page 199, just before the item pages themselves. Each quotation costs $20,000. We suggest that you start by picking a few components that may meet your needs, and then looking carefully at the price and the "description." *Note*: The description field often contains vital information about, for example:

- details about what features the component provides;
- compatibility or noncompatibility with other components;
- how many such components are needed in a wrist videophone (sometimes you need two or more!);

# Find Paths to Achieve the Targets

**TABLE E.1** Output for Module E—Find Paths to the Targets

Your Company's Name: _____

Final Set of Features to be Provided in Your Company's New Product: →

| Attribute and Level | Circle one number in each group |
|---|---|
| **Color:** | |
| Chrome | 1 |
| Black (standard) | 2 |
| Custom: brown, blue, other | 3 |
| Gold | 4 |
| **Display:** | |
| 1.5 inch diagonal, black & white | 1 |
| 1.5 inch diagonal, color | 2 |
| 2.0 inch diagonal, color | 3 |
| **Video Speed:** | |
| industry standard (2 frames/second) | 1 |
| standard videophone (12 frames/second) | 2 |
| computer movie quality (24 frames/second) | 3 |
| **Stored Numbers:** | |
| 4 numbers | 1 |
| 16 numbers | 2 |
| 32 numbers | 3 |
| **Talk Time:** | |
| 15-minute talk time, 10-hr. standby | 1 |
| 30-minute talk time, 16-hr. standby | 2 |
| 60-minute talk time, 24-hr. standby | 3 |
| **Messaging Service:** | |
| None | 1 |
| stores 2 minutes of voice messages | 2 |
| stores 5 minutes of voice messages | 3 |
| **Delivery:** | |
| within 24 hours | 1 |
| within 72 hours | 2 |

**Final Costs, Margins, Prices:**

| | New Product | | |
|---|---|---|---|
| | Year 2011 | Year 2012 | Year 2013 |
| COGS (without discounts) | $ | $ | $ |
| Volume Discount on Entire COGS | | | |
| Number of outside suppliers = | | | |
| If suppliers = 1, calculate 5% discount | | | |
| If suppliers = 2, calculate 2% discount | $ | $ | $ |
| If suppliers = 3 or 4, calculate 0% discount | | | |
| Systems with only one outside supplier | | | |
| If Controller, record undiscounted cost | $ | $ | $ |
| If Telephone, record undiscounted cost | $ | $ | $ |
| If Display, record undiscounted cost | $ | $ | $ |
| If Hsg. & Assy., record undiscounted cost | $ | $ | $ |
| Select most advantageous one (write name) | | | |
| If Controller, or Telephone, calculate 20% discount | | | |
| If Display, or Hsg. & Assy., calculate 10% discount | $ | $ | $ |
| If none have single o/s supplier, use 0% discount | | | |
| Net COGS (with discounts applied) | $ | $ | $ |
| Final Choice of Gross Margins | % | % | % |
| Final Choice of Prices | $ | $ | $ |

**Expenses:**

| | Amount |
|---|---|
| DEVELOPMENT (in 2011): | |
| Module A | $ K |
| Module B | $ K |
| Module C | $ K |
| Module D | $ K |
| Module E: | |
| number of Quotations looked at = | |
| multiply by $20,000 | $ K |
| Total Devel. Expense = | $ K |
| YEARLY (in 2011, 2012, 2013): | |
| Sum of Yearly Expenses for each Component = | $ M |

- combinations (sometimes a component does multiple functions, so you can skip buying one of the others).

You will find that you don't like some of your first choices (unsuitable, too expensive, etc.). So then you will have to continue looking at new quotations. You may have to make compromises in cost, features, etc. Don't be surprised if you have to look at 30, 40, or even more quotations until you have a satisfactory set of components that you can use.

## "SEQUENCE NUMBER"

In the real world, you often have to make successive attempts with one or more suppliers before you find the part that you can use. Sometimes the requirements aren't completely described, sometimes a vendor wants to lead you in a certain direction, and so on. Therefore we have established a rule that you must look at the quotations in a certain sequence. Within the menu for each type of component (bezel, RAM, battery, etc.) you will see a "sequence number" (= 1, 2, or 3) associated with each item. The rule for *each* type of component is:

- You must examine at least one "1" item before you examine any "2" item, and you must examine at least one "2" item before you examine any "3" item.

Within this rule, it does not matter which supplier is involved. For example, you can examine a "1" item from supplier "C," and then a "2" item from supplier "B."

## YOUR OUTPUT

Once you have made your component selections, you must record a number of items of final output, in Table E.1. This will comprise the following:

1. Your final decisions on product features, dictated by your component selections. These may or may not be the same as your original choices in Table B.2.
2. Your undiscounted COGS, any volume discounts that you earned, and your net COGS.
3. Your total expense in 2010 to develop this new product. This is the cost of all the information that you purchased. Add together the total expenses that you recorded at the lower left of Tables A.1, B.1, C.1, and D.1. Add to this the cost of all the quotations that you looked at (= number of quotations × $20,000).
4. The total yearly costs in 2011, 2012, and 2013 associated with using the components that you selected.
5. Your final decisions on prices and gross margins in 2010, 2012, and 2013. You may use those that you originally determined in Table C.1, or you may make last-minute changes.

## YOUR TASK FOR MODULE E

(Time limit: 60 minutes.) Purchase the information that you think you need. Select the components that will be used in *your company's* new wrist videophone. On the basis of your choices, determine your product's final set of features—and its costs (COGS), its final prices, and its gross margins in 2011, 2012, and 2013. Finally, record your development expenses and your subsequent yearly expenses.

## AFTER COMPLETING MODULE E

If you are in a classroom setting we strongly encourage the teams to share their final designs, costs, prices, and margins with each other. Either as

## Find Paths to Achieve the Targets

an individual reader or as a member of a class, depending on what information you obtained, here are some of the things you should have learned by the end Module E:

In Module A
- Your business scenario.
- *Your company's* product and name.
- Your business strategy.
- Last year's financial results.
- Market forecasts—sizes, shares.
- Competitive strengths and weaknesses.
- Competitors' costs.

In Module B
- Customer preferences.
- Market-feature table.
- Conjoint Analysis: buyer utilities.
- Product attributes: importance-value table.
- Features that will be in your product.

In Module C
- Historical and current prices for your current product.
- This year's projected financial results.
- Price trends.
- Price experience curve.
- Your planned gross margins.
- Cost experience curve.
- Target prices and costs (COGS) for 2011–2013.

In Module D
- Function–subsystem matrix.
- Current costs of subsystems and components.
- Cost targets for each subsystem.
- Subsystem cost gaps (2011–2013)

In Module E
- Final attributes or features of your new product.
- Final selection of components and suppliers.
- Volume and partnership discounts.
- COGS for new product (2011–2013).
- Prices for new product (2011–2013).
- Gross margins for new product (2011–2013).
- Development expenses (2010).
- Yearly expenses (2011–2013).

> You have completed Module E of the Exercise.
> Please return to Chapter 6, page 89.

# PRODUCT COST WORKSHEET

| Subsystem/Component | Targets for Critical Yr. = 201_ | 1st Iteration Quot. # | 1st Iteration Supplier | 1st Iteration Cost ($) | 2nd Iteration Quot. # | 2nd Iteration Supplier | 2nd Iteration Cost ($) | 3rd Iteration Quot. # | 3rd Iteration Supplier | 3rd Iteration Cost ($) | Final Quot. # | Final Supplier | Year 2011 | Year 2012 | Year 2013 | Yearly Expenses ($M) |
|---|---|---|---|---|---|---|---|---|---|---|---|---|---|---|---|---|
| **Display** | $ | | | $ | | | $ | | | $ | | | $ | $ | $ | |
| LCD | | | | $ | | | $ | | | $ | | | $ | $ | $ | $ M |
| Driver Circuit | | | | $ | | | $ | | | $ | | | $ | $ | $ | $ M |
| Bezel | | | | $ | | | $ | | | $ | | | $ | $ | $ | $ M |
| **Controller** | $ | | | $ | | | $ | | | $ | | | $ | $ | $ | |
| Processor | | | | $ | | | $ | | | $ | | | $ | $ | $ | $ M |
| RAM | | | | $ | | | $ | | | $ | | | $ | $ | $ | $ M |
| FLASH | | | | $ | | | $ | | | $ | | | $ | $ | $ | $ M |
| ASIC | | | | $ | | | $ | | | $ | | | $ | $ | $ | $ M |
| Timer Circuit | | | | $ | | | $ | | | $ | | | $ | $ | $ | $ M |
| **Camera** | $ | | | $ | | | $ | | | $ | | | $ | $ | $ | |
| Lens | | | | $ | | | $ | | | $ | | | $ | $ | $ | $ M |
| Image ASIC | | | | $ | | | $ | | | $ | | | $ | $ | $ | $ M |
| **Telephone** | $ | | | $ | | | $ | | | $ | | | $ | $ | $ | |
| Speaker | | | | $ | | | $ | | | $ | | | $ | $ | $ | $ M |
| Microphone | | | | $ | | | $ | | | $ | | | $ | $ | $ | $ M |
| Transmitter | | | | $ | | | $ | | | $ | | | $ | $ | $ | $ M |
| Receiver | | | | $ | | | $ | | | $ | | | $ | $ | $ | $ M |
| Power Amplifier | | | | $ | | | $ | | | $ | | | $ | $ | $ | $ M |
| **Power** | $ | | | $ | | | $ | | | $ | | | $ | $ | $ | |
| Batteries | | | | $ | | | $ | | | $ | | | $ | $ | $ | $ M |
| Rectifier Circuit | | | | $ | | | $ | | | $ | | | $ | $ | $ | $ M |
| **Housing & Assembly** | $ | | | $ | | | $ | | | $ | | | $ | $ | $ | |
| Housing & Strap | | | | $ | | | $ | | | $ | | | $ | $ | $ | $ M |
| Antenna | | | | $ | | | $ | | | $ | | | $ | $ | $ | $ M |
| Assembly | | | | $ | | | $ | | | $ | | | $ | $ | $ | $ M |
| **Total =** | $ | | | $ | | | $ | | | $ | | | $ | | | $ M |

ITERATIONS (for a critical year = 201_)

FINAL DECISION — Cost ($)

# Find Paths to Achieve the Targets

## INFORMATION MENU FOR MODULE E (PART 1): MENUS OF AVAILABLE QUOTATIONS FOR DISPLAY AND CONTROLLER

### DISPLAY Components

#### LCD:

| Item | Supplier | Component | Seq. # |
|---|---|---|---|
| 1 | In-House | 2" color LCD | 1 |
| 2 | A | 1.5" B&W LCD | 1 |
| 3 | A | 1.5" color LCD | 1 |
| 4 | B | 1.5" color LCD | 2 |
| 5 | B | 2" color LCD | 2 |
| 6 | C | 1.5" B&W LCD | 2 |
| 7 | C | 1.5" color LCD | 3 |
| 8 | C | 2" color LCD | 3 |

#### Driver Circuit:

| Item | Supplier | Component | Seq. # |
|---|---|---|---|
| 9 | In-House | Driver Circuit (movie) | 1 |
| 10 | A | Driver Circuit (std) | 1 |
| 11 | B | Driver Circuit (std) | 2 |
| 12 | B | Driver Circuit (video) | 1 |
| 13 | B | Driver Circuit (movie) | 2 |
| 14 | C | Driver Circuit (video) | 2 |

#### Bezel:

| Item | Supplier | Component | Seq. # |
|---|---|---|---|
| 15 | A | Bezel - chrome | 1 |
| 16 | A | Bezel - black | 2 |
| 17 | B | Bezel - black | 1 |
| 18 | C | Bezel - color | 1 |
| 19 | C | Bezel - gold plated | 1 |

### CONTROLLER Components

#### Processor:

| Item | Supplier | Component | Seq. # |
|---|---|---|---|
| 20 | In-House | Optimum 5006 Processor | 1 |
| 21 | A | Optimum 5006 Processor | 1 |
| 22 | B | Optimum 5006 Processor | 2 |
| 23 | B | Optimum 5012 Processor | 2 |
| 24 | C | Optimum 5012 Processor | 1 |
| 25 | C | Optimum 5012 Processor | 2 |

#### RAM:

| Item | Supplier | Component | Seq. # |
|---|---|---|---|
| 26 | A | 4K RAM | 1 |
| 27 | B | 4K RAM | 2 |
| 28 | B | 2K RAM | 1 |
| 29 | B | 8K SRAM | 3 |

#### FLASH:

| Item | Supplier | Component | Seq. # |
|---|---|---|---|
| 30 | A | 2K FLASH | 1 |
| 31 | A | 4K FLASH | 1 |
| 32 | B | 4K FLASH | 2 |
| 33 | C | 4K FLASH | 2 |
| 34 | A | 8K EPROM | 3 |

#### ASIC:

| Item | Supplier | Component | Seq. # |
|---|---|---|---|
| 35 | In-House | ASIC (24 fps) | 1 |
| 36 | A | ASIC (2 fps) | 2 |
| 37 | B | ASIC (2 fps) | 1 |
| 38 | B | ASIC (12 fps) | 2 |
| 39 | B | ASIC (24 fps) | 2 |
| 40 | C | ASIC (12 fps) | 1 |
| 41 | C | ASIC (24fps) | 3 |

#### Timer Circuit:

| Item | Supplier | Component | Seq. # |
|---|---|---|---|
| 42 | A | 600MHz Timer Circuit | 1 |
| 43 | B | 600MHz Timer Circuit | 2 |
| 44 | C | 600MHz Timer Circuit | 3 |
| 45 | A | 1200MHz Timer Circuit | 1 |
| 46 | B | 1200MHz Timer Circuit | 2 |

# INFORMATION MENU FOR MODULE E (PART 2): MENUS OF AVAILABLE QUOTATIONS FOR CAMERA AND TELEPHONE

## CAMERA Components

### Lens:

| Item | Supplier | Component | Seq. # |
|---|---|---|---|
| 47 | A | 10:1 Lens | 1 |
| 48 | B | 10:1 Lens | 3 |
| 49 | B | 20:1 Lens | 1 |
| 50 | C | 15:1 Lens | 2 |
| 51 | D | 20:1 Lens | 3 |

### Image ASIC:

| Item | Supplier | Component | Seq. # |
|---|---|---|---|
| 52 | In-House | HS Image ASIC | 1 |
| 53 | A | Image ASIC | 1 |
| 54 | B | Image ASIC | 2 |
| 55 | B | Steady Image ASIC | 3 |
| 56 | C | HS Image ASIC | 2 |

## TELEPHONE Components

### Speaker:

| Item | Supplier | Component | Seq. # |
|---|---|---|---|
| 57 | A | Speaker | 1 |
| 58 | B | Speaker | 2 |
| 59 | B | Speaker | 1 |
| 60 | C | Speaker | 1 |
| 61 | C | Speaker | 2 |

### Microphone:

| Item | Supplier | Component | Seq. # |
|---|---|---|---|
| 62 | A | Microphone | 1 |
| 63 | B | Microphone | 1 |
| 64 | C | Microphone | 2 |

### Transmitter:

| Item | Supplier | Component | Seq. # |
|---|---|---|---|
| 65 | A | Transmitter | 2 |
| 66 | B | Transmitter | 1 |
| 67 | B | Transmitter | 2 |
| 68 | C | Transmitter | 1 |

### Receiver:

| Item | Supplier | Component | Seq. # |
|---|---|---|---|
| 69 | A | Receiver | 3 |
| 70 | B | Receiver | 1 |
| 71 | C | Receiver | 2 |
| 72 | In-House | Receiver | 1 |
| 73 | C | Receiver | 3 |

### Power Amplifier:

| Item | Supplier | Component | Seq. # |
|---|---|---|---|
| 74 | A | Power Amplifier | 1 |
| 75 | B | Power Amplifier | 2 |
| 76 | C | Power Amplifier | 3 |

# Find Paths to Achieve the Targets

## INFORMATION MENU FOR MODULE E (PART 3): MENUS OF AVAILABLE QUOTATIONS FOR POWER AND HOUSING AND ASSEMBLY

### POWER Components

#### Battery:

| Item | Supplier | Component | Seq. # |
|---|---|---|---|
| 77 | A | NiCd Battery | 1 |
| 78 | B | NiCd Battery | 2 |
| 79 | B | NiMH Battery | 1 |
| 80 | C | NiCd Battery | 3 |
| 81 | C | NiMH Battery | 2 |
| 82 | C | Li-ion Battery | 3 |

#### Rectifier Circuit:

| Item | Supplier | Component | Seq. # |
|---|---|---|---|
| 83 | A | Rectifier Circuit | 1 |
| 84 | B | Rectifier Circuit | 2 |
| 85 | C | Rectifier Circuit | 1 |

### HOUSING & ASSEMBLY Components

#### Housing & Strap:

| Item | Supplier | Component | Seq. # |
|---|---|---|---|
| 86 | A | Housing & Strap - black | 1 |
| 87 | B | Housing & Strap - black | 2 |
| 88 | B | Housing & Strap - chrome | 1 |
| 89 | A | Housing & Strap - any color | 1 |
| 90 | B | Housing & Strap - gold plated | 1 |

#### Antenna:

| Item | Supplier | Component | Seq. # |
|---|---|---|---|
| 91 | A | Antenna | 1 |
| 92 | B | Antenna | 2 |
| 93 | C | Antenna | 1 |
| 94 | C | Antenna | 2 |

#### Assembly & Test:

| Item | Supplier | Component | Seq. # |
|---|---|---|---|
| 95 | In-House | Assembly & Test | 1 |
| 96 | C | Assembly & Test | 2 |
| 97 | D | Assembly & Test | 3 |

## Item A    Quotations for Display Components

**LCD:**

| Yes, I looked at this item. | Item | Seq. # | Supplier | Subsystem | Component | Description | 2010 | 2011 | 2012 | 2013 | Yearly Expense [$M] |
|---|---|---|---|---|---|---|---|---|---|---|
|  | 1 | 1 | In-House | Display | 2" color LCD | 2" color organic | $ 47.00 | $ 28.00 | $ 18.00 | $ 12.00 | $ 3.200 |

| Yes, I looked at this item. | Item | Seq. # | Supplier | Subsystem | Component | Description | 2010 | 2011 | 2012 | 2013 | Yearly Expense [$M] |
|---|---|---|---|---|---|---|---|---|---|---|
|  | 2 | 1 | A | Display | 1.5" B&W LCD | 1.5" diag. B&W | $ 10.00 | $ 9.50 | $ 9.00 | $ 8.50 | $ 0.800 |

| Yes, I looked at this item. | Item | Seq. # | Supplier | Subsystem | Component | Description | 2010 | 2011 | 2012 | 2013 | Yearly Expense [$M] |
|---|---|---|---|---|---|---|---|---|---|---|
|  | 3 | 1 | A | Display | 1.5" color LCD | 1.5" diag. color | $ 27.00 | $ 19.00 | $ 14.00 | $ 10.00 | $ 1.250 |

| Yes, I looked at this item. | Item | Seq. # | Supplier | Subsystem | Component | Description | 2010 | 2011 | 2012 | 2013 | Yearly Expense [$M] |
|---|---|---|---|---|---|---|---|---|---|---|
|  | 4 | 2 | B | Display | 1.5" color LCD | 3.5 cm color | $ 23.00 | $ 16.00 | $ 11.00 | $ 9.00 | $ 1.700 |

| Yes, I looked at this item. | Item | Seq. # | Supplier | Subsystem | Component | Description | 2010 | 2011 | 2012 | 2013 | Yearly Expense [$M] |
|---|---|---|---|---|---|---|---|---|---|---|
|  | 5 | 2 | B | Display | 2" color LCD | 5 cm color | $ 44.00 | $ 37.00 | $ 32.00 | $ 30.00 | $ 2.950 |

| Yes, I looked at this item. | Item | Seq. # | Supplier | Subsystem | Component | Description | 2010 | 2011 | 2012 | 2013 | Yearly Expense [$M] |
|---|---|---|---|---|---|---|---|---|---|---|
|  | 6 | 2 | C | Display | 1.5" B&W LCD | 1.5" diag. B&W | $ 8.50 | $ 8.00 | $ 7.00 | $ 6.50 | $ 0.400 |

| Yes, I looked at this item. | Item | Seq. # | Supplier | Subsystem | Component | Description | 2010 | 2011 | 2012 | 2013 | Yearly Expense [$M] |
|---|---|---|---|---|---|---|---|---|---|---|
|  | 7 | 3 | C | Display | 1.5" color LCD | 1.5" diag. color | $ 31.00 | $ 25.00 | $ 20.00 | $ 14.00 | $ 1.100 |

| Yes, I looked at this item. | Item | Seq. # | Supplier | Subsystem | Component | Description | 2010 | 2011 | 2012 | 2013 | Yearly Expense [$M] |
|---|---|---|---|---|---|---|---|---|---|---|
|  | 8 | 3 | C | Display | 2" color LCD | 2" diag. color | $ 58.00 | $ 49.00 | $ 39.00 | $ 29.00 | $ 1.300 |

**Driver Circuit:**

| Yes, I looked at this item. | Item | Seq. # | Supplier | Subsystem | Component | Description | 2010 | 2011 | 2012 | 2012 | Yearly Expense [$M] |
|---|---|---|---|---|---|---|---|---|---|---|
|  | 9 | 1 | In-House | Display | Driver Circuit (movie) | chip on glass (24 fps) | $ 25.00 | $ 20.00 | $ 18.00 | $ 14.00 | $ 0.850 |

| Yes, I looked at this item. | Item | Seq. # | Supplier | Subsystem | Component | Description | 2010 | 2011 | 2012 | 2013 | Yearly Expense [$M] |
|---|---|---|---|---|---|---|---|---|---|---|
|  | 10 | 1 | A | Display | Driver Circuit (std) | chip on board (2 fps) | $ 25.00 | $ 22.00 | $ 19.00 | $ 18.00 | $ 0.200 |

| Yes, I looked at this item. | Item | Seq. # | Supplier | Subsystem | Component | Description | 2010 | 2011 | 2012 | 2013 | Yearly Expense [$M] |
|---|---|---|---|---|---|---|---|---|---|---|
|  | 11 | 2 | B | Display | Driver Circuit (std) | chip on board (2 fps) | $ 25.00 | $ 16.00 | $ 12.00 | $ 10.00 | $ 0.200 |

| Yes, I looked at this item. | Item | Seq. # | Supplier | Subsystem | Component | Description | 2010 | 2011 | 2012 | 2013 | Yearly Expense [$M] |
|---|---|---|---|---|---|---|---|---|---|---|
|  | 12 | 1 | B | Display | Driver Circuit (video) | chip on board (12 fps) | $ 20.00 | $ 18.00 | $ 18.00 | $ 12.00 | $ 0.300 |

| Yes, I looked at this item. | Item | Seq. # | Supplier | Subsystem | Component | Description | 2010 | 2011 | 2012 | 2013 | Yearly Expense [$M] |
|---|---|---|---|---|---|---|---|---|---|---|

**Bezel:**

| Yes, I looked at this item. | Item | Seq. # | Supplier | Subsystem | Component | Description | 2010 | 2011 | 2012 | 2013 | Yearly Expense [$M] |
|---|---|---|---|---|---|---|---|---|---|---|
|  | 15 | 1 | A | Display | Bezel - chrome | chrome | $ 3.50 | $ 3.25 | $ 3.00 | $ 3.00 | $ 0.750 |

| Yes, I looked at this item. | Item | Seq. # | Supplier | Subsystem | Component | Description | 2010 | 2011 | 2012 | 2013 | Yearly Expense [$M] |
|---|---|---|---|---|---|---|---|---|---|---|
|  | 16 | 2 | A | Display | Bezel - black | black | $ 5.00 | $ 4.75 | $ 4.50 | $ 4.25 | $ 0.600 |

| Yes, I looked at this item. | Item | Seq. # | Supplier | Subsystem | Component | Description | 2010 | 2011 | 2012 | 2013 | Yearly Expense [$M] |
|---|---|---|---|---|---|---|---|---|---|---|
|  | 17 | 1 | B | Display | Bezel - black | black | $ 4.50 | $ 4.25 | $ 3.75 | $ 3.50 | $ 0.700 |

| Yes, I looked at this item. | Item | Seq. # | Supplier | Subsystem | Component | Description | 2010 | 2011 | 2012 | 2013 | Yearly Expense [$M] |
|---|---|---|---|---|---|---|---|---|---|---|
|  | 18 | 1 | C | Display | Bezel - color | color | $ 6.80 | $ 8.50 | $ 6.00 | $ 5.00 | $ 0.700 |

| Yes, I looked at this item. | Item | Seq. # | Supplier | Subsystem | Component | Description | 2010 | 2011 | 2012 | 2013 | Yearly Expense [$M] |
|---|---|---|---|---|---|---|---|---|---|---|
|  | 19 | 1 | C | Display | Bezel - gold plated | gold plated | $ 15.00 | $ 8.00 | $ 7.00 | $ 8.00 | $ 0.750 |

# Find Paths to Achieve the Targets

## Item. B1  Quotations for Controller Components (Part 1)

**Processor:**

| Supplier | Subsystem | Component | Description | 2010 | 2011 | 2012 | 2013 | Yearly Expense [$M] | Seq. # | Item | Yes, I looked at this item. |
|---|---|---|---|---|---|---|---|---|---|---|---|
| In-House | Controller | Optimum 5006 Processor | < 10 fps; for messages or stored numbers | $ 12.00 | $ 10.00 | $ 8.00 | $ 7.00 | $ 0.650 | 1 | 20 | |

| Supplier | Subsystem | Component | Description | 2010 | 2011 | 2012 | 2013 | Yearly Expense [$M] | Seq. # | Item | Yes, I looked at this item. |
|---|---|---|---|---|---|---|---|---|---|---|---|
| A | Controller | Optimum 5006 Processor | < 10 fps; no messages | $ 12.00 | $ 10.00 | $ 9.00 | $ 8.00 | $ 0.550 | 1 | 21 | |

| Supplier | Subsystem | Component | Description | 2010 | 2011 | 2012 | 2013 | Yearly Expense [$M] | Seq. # | Item | Yes, I looked at this item. |
|---|---|---|---|---|---|---|---|---|---|---|---|
| B | Controller | Optimum 5006 Processor | < 10 fps; no messages | N.A. | $ 7.50 | $ 6.75 | $ 6.25 | $ 1.000 | 2 | 22 | |

| Supplier | Subsystem | Component | Description | 2010 | 2011 | 2012 | 2013 | Yearly Expense [$M] | Seq. # | Item | Yes, I looked at this item. |
|---|---|---|---|---|---|---|---|---|---|---|---|
| B | Controller | Optimum 5012 Processor | > 10 fps; for messages or stored numbers | $ 10.00 | $ 8.00 | $ 7.00 | $ 6.50 | $ 0.450 | 2 | 23 | |

| Supplier | Subsystem | Component | Description | 2010 | 2011 | 2012 | 2013 | Yearly Expense [$M] | Seq. # | Item | Yes, I looked at this item. |
|---|---|---|---|---|---|---|---|---|---|---|---|
| C | Controller | Optimum 5012 Processor | > 10 fps; for messages or stored numbers | $ 12.00 | $ 9.00 | $ 8.00 | $ 7.00 | $ 0.600 | 1 | 24 | |

| Supplier | Subsystem | Component | Description | 2010 | 2011 | 2012 | 2013 | Yearly Expense [$M] | Seq. # | Item | Yes, I looked at this item. |
|---|---|---|---|---|---|---|---|---|---|---|---|
| C | Controller | Optimum 5012 Processor | > 10 fps; for messages or stored numbers | $ 15.00 | $ 10.00 | $ 7.00 | $ 5.00 | $ 0.950 | 2 | 25 | |

**RAM:**

| Supplier | Subsystem | Component | Description | 2010 | 2011 | 2012 | 2013 | Yearly Expense [$M] | Seq. # | Item | Yes, I looked at this item. |
|---|---|---|---|---|---|---|---|---|---|---|---|
| A | Controller | 4K RAM | for messages or stored numbers | $ 10.00 | $ 9.00 | $ 6.00 | $ 7.00 | $ 0.350 | 1 | 26 | |

| Supplier | Subsystem | Component | Description | 2010 | 2011 | 2012 | 2013 | Yearly Expense [$M] | Seq. # | Item | Yes, I looked at this item. |
|---|---|---|---|---|---|---|---|---|---|---|---|
| B | Controller | 4K RAM | for messages or stored numbers | $ 10.00 | $ 8.00 | $ 7.00 | $ 6.00 | $ 0.450 | 2 | 27 | |

| Supplier | Subsystem | Component | Description | 2010 | 2011 | 2012 | 2013 | Yearly Expense [$M] | Seq. # | Item | Yes, I looked at this item. |
|---|---|---|---|---|---|---|---|---|---|---|---|
| B | Controller | 2K RAM | for stored numbers only | $ 4.25 | $ 4.00 | $ 3.75 | $ 3.50 | $ 0.500 | 1 | 28 | |

| Supplier | Subsystem | Component | Description | 2010 | 2011 | 2012 | 2013 | Yearly Expense [$M] | Seq. # | Item | Yes, I looked at this item. |
|---|---|---|---|---|---|---|---|---|---|---|---|
| B | Controller | 8K SRAM | for msg's or stored #'s - no Flash needed | $ 20.00 | $ 16.00 | $ 12.00 | $ 10.00 | $ 1.150 | 3 | 29 | |

**FLASH:**

| Supplier | Subsystem | Component | Description | 2010 | 2011 | 2012 | 2013 | Yearly Expense [$M] | Seq. # | Item | Yes, I looked at this item. |
|---|---|---|---|---|---|---|---|---|---|---|---|
| A | Controller | 2K FLASH | for up to 16 #'s; no messages | $ 9.00 | $ 7.00 | $ 5.00 | $ 4.00 | $ 0.400 | 1 | 30 | |

| Supplier | Subsystem | Component | Description | 2010 | 2011 | 2012 | 2013 | Yearly Expense [$M] | Seq. # | Item | Yes, I looked at this item. |
|---|---|---|---|---|---|---|---|---|---|---|---|
| A | Controller | 4K FLASH | for up to 16 #'s and/or up to 5 min msg's | $ 20.00 | $ 18.00 | $ 16.00 | $ 14.00 | $ 0.400 | 1 | 31 | |

| Supplier | Subsystem | Component | Description | 2010 | 2011 | 2012 | 2013 | Yearly Expense [$M] | Seq. # | Item | Yes, I looked at this item. |
|---|---|---|---|---|---|---|---|---|---|---|---|
| B | Controller | 4K FLASH | for up to 32 #'s and/or up to 2 min msg's | $ 25.00 | $ 20.00 | $ 18.00 | $ 16.00 | $ 0.400 | 2 | 32 | |

| Supplier | Subsystem | Component | Description | 2010 | 2011 | 2012 | 2013 | Yearly Expense [$M] | Seq. # | Item | Yes, I looked at this item. |
|---|---|---|---|---|---|---|---|---|---|---|---|
| C | Controller | 4K FLASH | for up to 32 #'s and/or up to 2 min msg's | $ 24.00 | $ 20.00 | $ 19.00 | $ 18.00 | $ 0.450 | 2 | 33 | |

| Supplier | Subsystem | Component | Description | 2010 | 2011 | 2012 | 2013 | Yearly Expense [$M] | Seq. # | Item | Yes, I looked at this item. |
|---|---|---|---|---|---|---|---|---|---|---|---|
| A | Controller | 8K EPROM | 32 #'s and/or 5 min msg's - no RAM needed | $ 20.00 | $ 16.00 | $ 12.00 | $ 10.00 | $ 0.700 | 3 | 34 | |

Module E

## Item B2  Quotations for Controller Components (Part 2)

**ASIC:**

| | Item | Seq. # | Supplier | Subsystem | Component | Description | 2010 | 2011 | 2012 | 2013 | Yearly Expense [$M] |
|---|---|---|---|---|---|---|---|---|---|---|
| Yes, I looked at this item. | 35 | 1 | In-House | Controller | ASIC (24 fps) | chip on glass (24 fps) | $ 20.00 | $ 17.00 | $ 15.00 | $ 13.00 | $ 1.950 |
| Yes, I looked at this item. | 36 | 2 | A | Controller | ASIC (2 fps) | chip on board (2 fps) | $ 18.00 | $ 18.00 | $ 15.00 | $ 10.00 | $ 0.600 |
| Yes, I looked at this item. | 37 | 1 | B | Controller | ASIC (2 fps) | chip on board (2 fps) | $ 18.00 | $ 14.00 | $ 11.00 | $ 8.00 | $ 0.750 |
| Yes, I looked at this item. | 38 | 2 | B | Controller | ASIC (12 fps) | chip on board (12 fps) | $ 20.00 | $ 16.00 | $ 14.00 | $ 12.00 | $ 0.800 |
| Yes, I looked at this item. | 39 | 2 | B | Controller | ASIC (24 fps) | chip on glass (24 fps) | $ 25.00 | $ 20.00 | $ 16.00 | $ 14.00 | $ 1.550 |
| Yes, I looked at this item. | 40 | 1 | C | Controller | ASIC (12 fps) | chip on board (12 fps) | $ 16.00 | $ 12.00 | $ 10.00 | $ 8.00 | $ 1.700 |
| Yes, I looked at this item. | 41 | 3 | C | Controller | ASIC (24fps) | combo: display driver & controller ASIC on glass | N.A. | $ 25.00 | $ 20.00 | $ 15.00 | $ 2.100 |

**Timer Circuit:**

| | Item | Seq. # | Supplier | Subsystem | Component | Description | 2010 | 2011 | 2012 | 2013 | Yearly Expense [$M] |
|---|---|---|---|---|---|---|---|---|---|---|
| Yes, I looked at this item. | 42 | 1 | A | Controller | 600MHz Timer Circuit | 5006 processors only | $ 5.00 | $ 4.75 | $ 4.50 | $ 4.25 | $ 0.350 |
| Yes, I looked at this item. | 43 | 2 | B | Controller | 600MHz Timer Circuit | 5006 processors only | $ 5.00 | $ 4.50 | $ 4.25 | $ 3.75 | $ 0.750 |
| Yes, I looked at this item. | 44 | 3 | C | Controller | 600MHz Timer Circuit | 5006 processors only | $ 5.00 | $ 4.75 | $ 4.25 | $ 4.00 | $ 0.750 |
| Yes, I looked at this item. | 45 | 1 | A | Controller | 1200MHz Timer Circuit | for 5012 processors only | $ 6.00 | $ 5.00 | $ 4.50 | $ 4.00 | $ 0.800 |
| Yes, I looked at this item. | 46 | 2 | B | Controller | 1200MHz Timer Circuit | for 6012 processors only | $ 5.00 | $ 4.75 | $ 4.50 | $ 4.25 | $ 0.350 |

# Find Paths to Achieve the Targets

## Item C  Quotations for Camera Components

**Lens:**

| Supplier | Subsystem | Component | Description | 2010 | 2011 | 2012 | 2013 | Yearly Expense [$M] | Seq. # | Item | Yes, I looked at this item. |
|---|---|---|---|---|---|---|---|---|---|---|---|
| A | Camera | 10:1 Lens | standard | $ 13.00 | $ 12.00 | $ 11.00 | $ 10.00 | $ 0.750 | 1 | 47 | |

| Supplier | Subsystem | Component | Description | 2010 | 2011 | 2012 | 2013 | Yearly Expense [$M] | Seq. # | Item | Yes, I looked at this item. |
|---|---|---|---|---|---|---|---|---|---|---|---|
| B | Camera | 10:1 Lens | standard | $ 13.00 | $ 11.00 | $ 8.00 | $ 6.00 | $ 0.750 | 3 | 48 | |

| Supplier | Subsystem | Component | Description | 2010 | 2011 | 2012 | 2013 | Yearly Expense [$M] | Seq. # | Item | Yes, I looked at this item. |
|---|---|---|---|---|---|---|---|---|---|---|---|
| B | Camera | 20:1 Lens | high resolution | $ 15.00 | $ 10.00 | $ 9.00 | $ 8.00 | $ 0.750 | 1 | 49 | |

| Supplier | Subsystem | Component | Description | 2010 | 2011 | 2012 | 2013 | Yearly Expense [$M] | Seq. # | Item | Yes, I looked at this item. |
|---|---|---|---|---|---|---|---|---|---|---|---|
| C | Camera | 15:1 Lens | super clear | $ 13.00 | $ 10.00 | $ 9.00 | $ 8.00 | $ 0.550 | 2 | 50 | |

| Supplier | Subsystem | Component | Description | 2010 | 2011 | 2012 | 2013 | Yearly Expense [$M] | Seq. # | Item | Yes, I looked at this item. |
|---|---|---|---|---|---|---|---|---|---|---|---|
| D | Camera | 20:1 Lens | high resolution | $ 16.00 | $ 10.00 | $ 8.00 | $ 7.00 | $ 0.400 | 3 | 51 | |

**Image ASIC:**

| Supplier | Subsystem | Component | Description | 2010 | 2011 | 2012 | 2013 | Yearly Expense [$M] | Seq. # | Item | Yes, I looked at this item. |
|---|---|---|---|---|---|---|---|---|---|---|---|
| In-house | Camera | HS Image ASIC | for >10 fps | $ 35.00 | $ 30.00 | $ 25.00 | $ 20.00 | $ 1.500 | 1 | 52 | |

| Supplier | Subsystem | Component | Description | 2010 | 2011 | 2012 | 2013 | Yearly Expense [$M] | Seq. # | Item | Yes, I looked at this item. |
|---|---|---|---|---|---|---|---|---|---|---|---|
| A | Camera | Image ASIC | for <10 fps | $ 22.00 | $ 20.00 | $ 19.00 | $ 18.00 | $ 0.750 | 1 | 53 | |

| Supplier | Subsystem | Component | Description | 2010 | 2011 | 2012 | 2013 | Yearly Expense [$M] | Seq. # | Item | Yes, I looked at this item. |
|---|---|---|---|---|---|---|---|---|---|---|---|
| B | Camera | Image ASIC | for <10 fps | $ 22.00 | $ 18.00 | $ 15.00 | $ 14.00 | $ 0.750 | 2 | 54 | |

| Supplier | Subsystem | Component | Description | 2010 | 2011 | 2012 | 2013 | Yearly Expense [$M] | Seq. # | Item | Yes, I looked at this item. |
|---|---|---|---|---|---|---|---|---|---|---|---|
| B | Camera | Steady Image ASIC | for <10 fps, steady image | $ 22.00 | $ 20.00 | $ 18.00 | $ 16.00 | $ 0.950 | 3 | 55 | |

| Supplier | Subsystem | Component | Description | 2010 | 2011 | 2012 | 2013 | Yearly Expense [$M] | Seq. # | Item | Yes, I looked at this item. |
|---|---|---|---|---|---|---|---|---|---|---|---|
| C | Camera | HS Image ASIC | for >10 fps | $ 28.00 | $ 23.00 | $ 19.00 | $ 15.00 | $ 2.250 | 2 | 56 | |

# Item D1  Quotations for Telephone Components (Part 1)

**Speaker:**

| Yes, I looked at this item. | Item | Seq. | Supplier | Subsystem | Component | Description | 2010 | 2011 | 2012 | 2013 | Yearly Expense [$M] |
|---|---|---|---|---|---|---|---|---|---|---|
| | 57 | 1 | A | Telephone | Speaker | 2.5 MOS voice quality | $ 5.00 | $ 5.00 | $ 4.75 | $ 4.50 | $ 0.550 |
| | 58 | 2 | B | Telephone | Speaker | 2.5 MOS voice quality | $ 5.00 | $ 4.75 | $ 4.50 | $ 4.35 | $ 0.550 |
| | 59 | 1 | B | Telephone | Speaker | 3 MOS voice quality | $ 5.50 | $ 5.25 | $ 5.00 | $ 4.75 | $ 0.600 |
| | 60 | 1 | C | Telephone | Speaker | 3 MOS voice quality | $ 5.25 | $ 5.00 | $ 4.75 | $ 4.50 | $ 0.600 |
| | 61 | 2 | C | Telephone | Speaker | combination - includes microphone | $ 8.00 | $ 6.00 | $ 5.00 | $ 4.00 | $ 0.850 |

**Microphone:**

| Yes, I looked at this item. | Item | Seq. | Supplier | Subsystem | Component | Description | 2010 | 2011 | 2012 | 2013 | Yearly Expense [$M] |
|---|---|---|---|---|---|---|---|---|---|---|
| | 62 | 1 | A | Telephone | Microphone | water proof | $ 3.00 | $ 2.75 | $ 2.75 | $ 2.50 | $ 0.550 |
| | 63 | 1 | B | Telephone | Microphone | water resistant | $ 2.75 | $ 2.50 | $ 2.25 | $ 2.35 | $ 0.550 |
| | 64 | 2 | C | Telephone | Microphone | combination - includes speaker | $ 8.00 | $ 6.00 | $ 5.00 | $ 4.00 | $ 0.850 |

**Transmitter:**

| Yes, I looked at this item. | Item | Seq. | Supplier | Subsystem | Component | Description | 2010 | 2011 | 2012 | 2013 | Yearly Expense [$M] |
|---|---|---|---|---|---|---|---|---|---|---|
| | 65 | 2 | A | Telephone | Transmitter | 3 devices needed to provide transmitter function | $ 10.00 | $ 9.00 | $ 8.00 | $ 7.00 | $ 0.600 |
| | 66 | 1 | B | Telephone | Transmitter | 2 devices needed to provide transmitter function | $ 10.00 | $ 8.00 | $ 7.00 | $ 6.00 | $ 0.600 |
| | 67 | 2 | B | Telephone | Transmitter | 1 device provides transmitter function | $ 12.00 | $ 9.00 | $ 8.00 | $ 7.00 | $ 0.550 |
| | 68 | 1 | C | Telephone | Transmitter | combination transmitter & receiver chip | $ 35.00 | $ 21.00 | $ 16.00 | $ 11.00 | $ 1.600 |

# Find Paths to Achieve the Targets

## Item D2 Quotations for Telephone Components (Part 2)

**Receiver:**

| Supplier | Subsystem | Component | Description | 2010 | 2011 | 2012 | 2013 | Yearly Expense [$M] | Seq. # | Item | Yes, I looked at this item. |
|---|---|---|---|---|---|---|---|---|---|---|---|
| A | Telephone | Receiver | 3 devices for receiver function | $ 15.00 | $ 14.00 | $ 13.00 | $ 12.00 | $ 0.550 | 3 | 69 | |

| Supplier | Subsystem | Component | Description | 2010 | 2011 | 2012 | 2013 | Yearly Expense [$M] | Seq. # | Item | Yes, I looked at this item. |
|---|---|---|---|---|---|---|---|---|---|---|---|
| B | Telephone | Receiver | 2 devices for receiver function | $ 16.00 | $ 13.00 | $ 11.00 | $ 10.00 | $ 0.600 | 1 | 70 | |

| Supplier | Subsystem | Component | Description | 2010 | 2011 | 2012 | 2013 | Yearly Expense [$M] | Seq. # | Item | Yes, I looked at this item. |
|---|---|---|---|---|---|---|---|---|---|---|---|
| C | Telephone | Receiver | 1 device for receiver function | $ 18.00 | $ 13.00 | $ 12.00 | $ 10.00 | $ 0.600 | 2 | 71 | |

| Supplier | Subsystem | Component | Description | 2010 | 2011 | 2012 | 2013 | Yearly Expense [$M] | Seq. # | Item | Yes, I looked at this item. |
|---|---|---|---|---|---|---|---|---|---|---|---|
| In-House | Telephone | Receiver | combination transmitter & receiver chip | $ 30.00 | $ 18.00 | $ 15.00 | $ 12.00 | $ 1.750 | 1 | 72 | |

| Supplier | Subsystem | Component | Description | 2010 | 2011 | 2012 | 2013 | Yearly Expense [$M] | Seq. # | Item | Yes, I looked at this item. |
|---|---|---|---|---|---|---|---|---|---|---|---|
| C | Telephone | Receiver | combination transmitter & receiver chip | $ 35.00 | $ 21.00 | $ 16.00 | $ 11.00 | $ 1.600 | 3 | 73 | |

**Power Amplifier:**

| Supplier | Subsystem | Component | Description | 2010 | 2011 | 2012 | 2013 | Yearly Expense [$M] | Seq. # | Item | Yes, I looked at this item. |
|---|---|---|---|---|---|---|---|---|---|---|---|
| A | Telephone | Power Amplifier | standard efficiency | $ 12.00 | $ 10.00 | $ 9.00 | $ 8.00 | $ 0.550 | 1 | 74 | |

| Supplier | Subsystem | Component | Description | 2010 | 2011 | 2012 | 2013 | Yearly Expense [$M] | Seq. # | Item | Yes, I looked at this item. |
|---|---|---|---|---|---|---|---|---|---|---|---|
| B | Telephone | Power Amplifier | high efficiency | $ 10.00 | $ 9.00 | $ 8.00 | $ 7.00 | $ 0.550 | 2 | 75 | |

| Supplier | Subsystem | Component | Description | 2010 | 2011 | 2012 | 2013 | Yearly Expense [$M] | Seq. # | Item | Yes, I looked at this item. |
|---|---|---|---|---|---|---|---|---|---|---|---|
| C | Telephone | Power Amplifier | high efficiency | $ 9.00 | $ 9.00 | $ 8.00 | $ 7.00 | $ 0.550 | 3 | 76 | |

# Item E  Quotations for Power Components

**Battery:**

| Yes, I looked at this item. | Item | Seq. # | Supplier | Subsystem | Component | Description | 2010 | 2011 | 2012 | 2013 | Yearly Expense [$M] |
|---|---|---|---|---|---|---|---|---|---|---|
|  | 77 | 1 | A | Power | NiCd Battery | NiCd - 15 min talk time | $ 12.00 | $ 11.00 | $ 10.00 | $ 9.00 | $ 0.550 |

| Yes, I looked at this item. | Item | Seq. # | Supplier | Subsystem | Component | Description | 2010 | 2011 | 2012 | 2013 | Yearly Expense [$M] |
|---|---|---|---|---|---|---|---|---|---|---|
|  | 78 | 2 | B | Power | NiCd Battery | NiCd - 15 min talk time | $ 12.00 | $ 10.00 | $ 9.00 | $ 8.00 | $ 0.550 |

| Yes, I looked at this item. | Item | Seq. # | Supplier | Subsystem | Component | Description | 2010 | 2011 | 2012 | 2013 | Yearly Expense [$M] |
|---|---|---|---|---|---|---|---|---|---|---|
|  | 79 | 1 | B | Power | NiMH Battery | NiMH - 30 min talk time | $ 15.00 | $ 12.00 | $ 10.00 | $ 9.00 | $ 1.000 |

| Yes, I looked at this item. | Item | Seq. # | Supplier | Subsystem | Component | Description | 2010 | 2011 | 2012 | 2013 | Yearly Expense [$M] |
|---|---|---|---|---|---|---|---|---|---|---|
|  | 80 | 3 | C | Power | NiCd Battery | NiCd - 15 min talk time | $ 12.00 | $ 11.00 | $ 10.00 | $ 9.00 | $ 0.550 |

| Yes, I looked at this item. | Item | Seq. # | Supplier | Subsystem | Component | Description | 2010 | 2011 | 2012 | 2013 | Yearly Expense [$M] |
|---|---|---|---|---|---|---|---|---|---|---|
|  | 81 | 2 | C | Power | NiMH Battery | NiMH - 30 min talk time | $ 15.00 | $ 13.00 | $ 12.00 | $ 10.00 | $ 1.000 |

| Yes, I looked at this item. | Item | Seq. # | Supplier | Subsystem | Component | Description | 2010 | 2011 | 2012 | 2013 | Yearly Expense [$M] |
|---|---|---|---|---|---|---|---|---|---|---|
|  | 82 | 3 | C | Power | Li-ion Battery | combination – 30 min talk time, battery & strap | $ 14.00 | $ 13.00 | $ 11.00 | $ 10.00 | $ 1.250 |

**Rectifier Circuit:**

| Yes, I looked at this item. | Item | Seq. # | Supplier | Subsystem | Component | Description | 2010 | 2011 | 2012 | 2013 | Yearly Expense [$M] |
|---|---|---|---|---|---|---|---|---|---|---|
|  | 83 | 1 | A | Power | Rectifier Circuit | 2 devices needed to provide rectifier function | $ 8.00 | $ 7.00 | $ 6.00 | $ 5.00 | $ 0.550 |

| Yes, I looked at this item. | Item | Seq. # | Supplier | Subsystem | Component | Description | 2010 | 2011 | 2012 | 2013 | Yearly Expense [$M] |
|---|---|---|---|---|---|---|---|---|---|---|
|  | 84 | 2 | B | Power | Rectifier Circuit | 1 device needed to provide rectifier function | $ 8.00 | $ 6.00 | $ 5.00 | $ 5.00 | $ 0.750 |

| Yes, I looked at this item. | Item | Seq. # | Supplier | Subsystem | Component | Description | 2010 | 2011 | 2012 | 2013 | Yearly Expense [$M] |
|---|---|---|---|---|---|---|---|---|---|---|
|  | 85 | 1 | C | Power | Rectifier Circuit | 1 device needed to provide rectifier function | $ 8.00 | $ 7.00 | $ 6.00 | $ 5.00 | $ 0.550 |

# Find Paths to Achieve the Targets

## Item F  Quotations for Housing and Assembly Components

### Housing & Strap:

| Supplier | Subsystem | Component | Description | 2010 | 2011 | 2012 | 2013 | Yearly Expense ($M) | Seq. # | Item | Yes, I looked at this item |
|---|---|---|---|---|---|---|---|---|---|---|---|
| A | Housing & Asy. | Housing & Strap - black | black | $ 1.00 | $ 1.00 | $ 1.00 | $ 1.00 | $ 0.500 | 1 | 86 | |

| Supplier | Subsystem | Component | Description | 2010 | 2011 | 2012 | 2013 | Yearly Expense ($M) | Seq. # | Item | Yes, I looked at this item |
|---|---|---|---|---|---|---|---|---|---|---|---|
| B | Housing & Asy. | Housing & Strap - black | black | $ 1.00 | $ 1.00 | $ 1.00 | $ 1.00 | $ 0.750 | 2 | 87 | |

| Supplier | Subsystem | Component | Description | 2010 | 2011 | 2012 | 2013 | Yearly Expense ($M) | Seq. # | Item | Yes, I looked at this item |
|---|---|---|---|---|---|---|---|---|---|---|---|
| B | Housing & Asy. | Housing & Strap - chrome | chrome | $ 2.00 | $ 2.00 | $ 2.00 | $ 2.00 | $ 0.525 | 1 | 88 | |

| Supplier | Subsystem | Component | Description | 2010 | 2011 | 2012 | 2013 | Yearly Expense ($M) | Seq. # | Item | Yes, I looked at this item |
|---|---|---|---|---|---|---|---|---|---|---|---|
| A | Housing & Asy. | Housing & Strap - any color | any color | $ 6.75 | $ 6.50 | $ 6.25 | $ 6.00 | $ 0.550 | 1 | 89 | |

| Supplier | Subsystem | Component | Description | 2010 | 2011 | 2012 | 2013 | Yearly Expense ($M) | Seq. # | Item | Yes, I looked at this item |
|---|---|---|---|---|---|---|---|---|---|---|---|
| B | Housing & Asy. | Housing & Strap - gold plated | gold plated | $ 15.00 | $ 14.00 | $ 14.00 | $ 13.00 | $ 0.550 | 1 | 90 | |

### Antenna:

| Supplier | Subsystem | Component | Description | 2010 | 2011 | 2012 | 2013 | Yearly Expense ($M) | Seq. # | Item | Yes, I looked at this item |
|---|---|---|---|---|---|---|---|---|---|---|---|
| A | Housing & Asy. | Antenna | straight antenna | $ 4.00 | $ 3.75 | $ 3.50 | $ 3.25 | $ 0.550 | 1 | 91 | |

| Supplier | Subsystem | Component | Description | 2010 | 2011 | 2012 | 2013 | Yearly Expense ($M) | Seq. # | Item | Yes, I looked at this item |
|---|---|---|---|---|---|---|---|---|---|---|---|
| B | Housing & Asy. | Antenna | coil antenna | $ 3.50 | $ 3.25 | $ 3.00 | $ 3.00 | $ 0.600 | 2 | 92 | |

| Supplier | Subsystem | Component | Description | 2010 | 2011 | 2012 | 2013 | Yearly Expense ($M) | Seq. # | Item | Yes, I looked at this item |
|---|---|---|---|---|---|---|---|---|---|---|---|
| C | Housing & Asy. | Antenna | combination -- includes black strap | $ 3.75 | $ 3.50 | $ 3.25 | $ 3.00 | $ 0.750 | 1 | 93 | |

| Supplier | Subsystem | Component | Description | 2010 | 2011 | 2012 | 2013 | Yearly Expense ($M) | Seq. # | Item | Yes, I looked at this item |
|---|---|---|---|---|---|---|---|---|---|---|---|
| C | Housing & Asy. | Antenna | combination -- includes color strap | $ 6.00 | $ 5.00 | $ 4.00 | $ 3.50 | $ 0.800 | 2 | 94 | |

### Assembly & Test:

| Supplier | Subsystem | Component | Description | 2010 | 2011 | 2012 | 2013 | Yearly Expense ($M) | Seq. # | Item | Yes, I looked at this item |
|---|---|---|---|---|---|---|---|---|---|---|---|
| In-House | Housing & Asy. | Assembly & Test | current in-house assembly line | $ 10.00 | $ 9.50 | $ 9.00 | $ 8.00 | $ 1.450 | 1 | 95 | |

| Supplier | Subsystem | Component | Description | 2010 | 2011 | 2012 | 2013 | Yearly Expense ($M) | Seq. # | Item | Yes, I looked at this item |
|---|---|---|---|---|---|---|---|---|---|---|---|
| C | Housing & Asy. | Assembly & Test | contract manufacturer | $ 10.00 | $ 9.50 | $ 9.00 | $ 8.00 | $ 1.600 | 2 | 96 | |

| Supplier | Subsystem | Component | Description | 2010 | 2011 | 2012 | 2013 | Yearly Expense ($M) | Seq. # | Item | Yes, I looked at this item |
|---|---|---|---|---|---|---|---|---|---|---|---|
| D | Housing & Asy. | Assembly & Test | contract manufacturer | $ 11.50 | $ 9.00 | $ 7.50 | $ 7.50 | $ 1.400 | 3 | 97 | |

# Module F

## Get Financial Results

In this module you will learn how well you did in the Exercise. You launched *your company's* new product into the competitive marketplace at the beginning of 2011. Assume that it is now three years later, at the end of 2013.

**YOUR OBJECTIVE**

(Time limit: 30 minutes.) Determine the market share that *your company's* new product captured in 2011, 2012, and 2013. Then determine your resulting revenues in those years. Subtract your COGS and other expenses in those years to determine your net income. Add the net income from 2010 (provided), 2011, 2012, and 2013 to get your four-year total net income and that is your financial result.

**COMPUTER SIMULATION METHOD**

If you use this book in a class where the instructor uses Sawtooth's ACA® Market Module, then it is a simple matter to enter the data pertaining to *your company's* new product and let the simulation run, delivering your results in a few minutes. The simulation puts your old and new products into a dynamic marketplace—along with products from three adaptable and aggressive competitors. It analyzes buyer utilities and behavior, and produces the market share for all five products in the years 2011, 2012, and 2013. These market shares then are applied to the addressable markets, and financial results are calculated.

## MANUAL METHOD

However, let us assume that you don't have access to the computer simulation. In that case you will have to use the two worksheets provided in this module. Obviously we can't reproduce all the stochastic factors that the computer simulation provides, but we have set up the worksheet so that you will get comparable results—without being overburdened with tedious calculations. First you have to calculate your market share, then you have to calculate your consequent financial results. It will take you about a half-hour to do this task.

## MARKET SHARE WORKSHEET

| Attribute, and Level | Put a "1" next to the level you chose | GOVERNMENT Weighted Utility | GOVERNMENT Score (pick the Wtd. Utility that corresponds to your choice) | BUSINESS Weighted Utility | BUSINESS Score (pick the Wtd. Utility that corresponds to your choice) | CONSUMER Weighted Utility | CONSUMER Score (pick the Wtd. Utility that corresponds to your choice) | Product's Market Share (units) |
|---|---|---|---|---|---|---|---|---|
| **Color:** | | | | | | | | |
| Chroma | - | 1.4 | | 6.2 | | 2.7 | | |
| Black (standard) | - | 12.3 | | 0.8 | | 0.6 | | |
| Custom: brown, blue, other | - | 2.1 | | 2.8 | | 2.9 | | |
| Gold | - | 0.2 | | 0.6 | | 1.0 | | |
| **Display:** | | | | | | | | |
| 1.5 inch diagonal, black & white | - | 1.3 | | 2.9 | | 3.2 | | |
| 1.5 inch diagonal, color | - | 10.3 | | 9.7 | | 1.8 | | |
| 2.0 inch diagonal, color | - | 8.3 | | 1.9 | | 0.3 | | |
| **Video Speed:** | | | | | | | | |
| industry standard (2 frames/second) | - | 0.0 | | 7.3 | | 7.0 | | |
| standard video phone (12 frames/second) | - | 4.8 | | 8.6 | | 3.2 | | |
| computer movie quality (24 frames/second) | - | 5.3 | | 0.2 | | 0.0 | | |
| **Stored Numbers:** | | | | | | | | |
| 4 numbers | - | 0.0 | | 1.6 | | 1.7 | | |
| 16 numbers | - | 5.0 | | 3.0 | | 0.9 | | |
| 32 numbers | - | 1.7 | | 1.8 | | 0.0 | | |
| **Talk Time:** | | | | | | | | |
| 15-minute talk time, 10-hr. standby | - | 0.0 | | 1.7 | | 6.5 | | |
| 30-minute talk time, 16-hr. standby | - | 12.2 | | 3.1 | | 4.1 | | |
| 60-minute talk time, 24-hr. standby | - | 11.3 | | 1.6 | | 0.2 | | |
| **Messaging Service:** | | | | | | | | |
| None | - | 1.5 | | 0.4 | | 4.7 | | |
| stores 2 minutes of voice messages | - | 0.6 | | 0.4 | | 1.6 | | |
| stores 5 minutes of voice messages | - | 0.7 | | 0.3 | | 0.0 | | |
| **Delivery:** | | | | | | | | |
| within 24 hours | - | 0.0 | | 0.2 | | 0.0 | | |
| within 72 hours | - | 1.3 | | 0.4 | | 0.4 | | |
| Score subtotal = | | | | | | | | |
| **Price** | | | | | | | | |
| $150 | | 11.5 | | 36.5 | | 73.2 | | |
| $250 | | 12.7 | | 30.3 | | 49.7 | | |
| $350 | | 8.5 | | 17.9 | | 16.2 | | |
| $450 | | 5.5 | | 9.8 | | 2.8 | | |
| $550 | | 0.2 | | 1.4 | | 0.0 | | |
| Your Price in 2011 = | $ - | | | | | | | |
| Your Price in 2012 = | $ - | | | | | | | |
| Your Price in 2013 = | $ - | | | | | | | If any number calculates to be less than zero, enter zero. |
| | | Share of Segment | Total Score | Share of Segment | Total Score | Share of Segment | Total Score | |
| Total Score for 2011 | | % | - | % | - | % | - | |
| Total Score for 2012 | | % | - | % | - | % | - | |
| Total Score for 2013 | | % | - | % | - | % | - | |

| | Total Market/Segment | Sales in Segment | Total Market Segment | Sales in Segment | Total Market Segment | Sales in Segment | Total Market (units) | Total Units Sold | Units Market Share |
|---|---|---|---|---|---|---|---|---|---|
| Unit Sales in 2011 | 400,000 | - | 200,000 | - | 150,000 | - | 750,000 | - | % |
| Unit Sales in 2012 | 500,000 | - | 600,000 | - | 500,000 | - | 1,600,000 | - | % |
| Unit Sales in 2013 | 600,000 | - | 1,200,000 | - | 1,600,000 | - | 3,400,000 | - | % |

| | Score for Price in a Year = A - B*(Price for that Year) | | | | | | Share of Segment = = (1.3%)*(Total Score) - 20% |
|---|---|---|---|---|---|---|---|
| | GOVERNMENT | | BUSINESS | | CONSUMER | | |
| | A | B | A | B | A | B | |
| year 2011: | 18.9 | 0.034 | 49.5 | 0.090 | 105.0 | 0.233 | |
| year 2012: | 16.0 | 0.029 | 39.6 | 0.072 | 78.8 | 0.175 | |
| year 2013: | 13.6 | 0.025 | 31.7 | 0.058 | 59.1 | 0.131 | |

## CALCULATING MARKET SHARE

On the Market Share Worksheet:

- For each of the first seven product attributes, indicate the level you chose to offer in your product. (Refer to your output in Table E.1, page 195.)
- For each of the market segments, write the "score" that corresponds to the "weighted utility" for the level you chose. For example, if you chose a 1.5-inch diagonal color display, the scores for Government, Business and Consumer would be 10.3, 9.7, and 1.8, respectively. (The weighted utilities have been calculated to reflect the relative importance of the attributes and the buyer utilities of the levels.)
- Add up the subtotal of the scores for the first seven attributes, for all three market segments.
- Enter the prices of *your company's* new product in 2011, 2012, and 2013.
- The scores for your price depend on the market segment, since each segment has a different set of buyer utilities and relative importance for price. Moreover, each segment expects a different price erosion or experience curve. We ask that you calculate the scores for each of the three years and for all three segments according to the equation: (Price score) = A − (B × Price), where A and B are constants as shown in the table.
- Add the subtotal scores to the yearly price scores to get total scores for each of the years 2011, 2012, and 2013. Do this for all three market segments.
- Then, use the equation: (Segment share) = {1.3% × (Total score) − 20%} to get the share of market segment captured by the new product in each year.
- Then, the sales in the segment in each of the years is simply = (Segment share) × (Total market segment)
- Add across and get your total units sold in each year. Divide by the total market units to get your final market share of all the units you sold, in all three years. You will take these three numbers into the next worksheet.

## CALCULATING FINANCIAL RESULTS

On the Financial Results Worksheet:

- First, enter the market share (%) for your new product in each of the years 2011, 2012, and 2013. Then, enter your new product's per-unit prices and costs (COGS) for those years. (Refer to your output in Table E.1.)

Module F

## FINANCIAL RESULTS WORKSHEET

Put in *Your Company's* new product's Market Share (%), Price (ea.), COGS (ea.) and Yearly Expenses ($M).

= (Old Product Share) + (New Product Share)

Revenue = (New Prod. Share) * (Tot. Mkt. Size) * (unit Price)
COGS = (New Prod. Share) * (Tot. Mkt. Size) * (unit COGS)

| Your Company's Name: | Year 2010 Old Product = Total | Year 2011 Old Product | Year 2011 New Product | Year 2011 Total | Year 2012 Old Product | Year 2012 New Product | Year 2012 Total | Year 2013 Old Product | Year 2013 New Product | Year 2013 Total | 2010-2013 Grand Total Net Income |
|---|---|---|---|---|---|---|---|---|---|---|---|
| Total Market Size (units) | 425,000 | | | 750,000 | | | 1,600,000 | | | 3,400,000 | |
| Old Product's Share (%) | 37% | 10% | | % | 13% | | % | 16% | | % | |
| New Product's Share (%) | | | | | | | | | % | | |
| Unit PRICE | $ 525.00 | $ 475.00 | $ | | $ 408.00 | $ | | $ 343.00 | $ | | |
| Unit COGS | $ 220.00 | $ 99.00 | $ | | $ 182.00 | $ | | $ 160.00 | $ | | |
| Revenue ($M) | $ 102.0 | $ 35.6 | $ | M $ | $ 84.9 | $ | M $ | $ 186.6 | M $ | M $ | |
| COGS ($M) | $ 42.8 | $ 14.9 | $ | $ | $ 37.9 | $ | $ | $ 87.0 | M $ | M $ | |
| Gross Margin ($M) | 59.2 | | | M | | | M | | | M | |
| Gross Margin (%) | 58% | | | % | | | % | | % | % | |
| Direct Expenses ($M) (incl. 2010 Devel. Expenses =)| $ | $ 22.0 | $ | $ | $ 42.0 | $ | $ | $ 74.0 | M $ | $ | |
| Allocated Expenses ($M) | 7.4 | | | M | | | M | | | M | |
| NET INCOME ($M) | $ | | | $ M | | | $ M | | | $ M | $ M |
| ROS % | % | | | % | | | % | | | % | Total Net Income |

Put in your Development Expense ($M) for 2010.

= $24M + (Devel. Expense)

= (Gross Margin) - (Direct Exp.) - (Allocated Exp.)

= (Net Income) / (Revenue)

= Revenue - COGS

= (G.M.) / Revenue

= 0.28 * Direct Expenses

# Get Financial Results

- Calculate your new product's revenues and costs for each of the three years: Revenue = share × (Market size) × (Unit price). COGS = share × (Market size) × (Unit COGS). Then add these numbers to their counterparts for the old product to get your total revenues and COGS in all three years.
- For each of 2011, 2012, and 2013, calculate your net gross margin in $million ( = Revenue − COGS), and in % ( = Margin/Revenues).
- Enter your development expenses that you incurred in 2010, and the annual expenses associated with the components used in the new product in 2011, 2012, and 2013. (Refer to your output in Table E.1.)
- We have provided representative quantities for some of the non-COGS direct expenses that you incur in all four years. In many companies these costs are determined by tracking the actual consumption of resources by each product, in others they are merely totaled and allocated to products based upon product revenues, and in others they are determined by something in between. The simulation calculates them based upon certain consumption rates, but to make things easy for the reader doing the manual calculation, we simply provide numbers that are representative. Add $24 million to year 2010 development expense to get your 2010 total direct expenses. Then add the figure you have for your annual expenses to the numbers we provided to get your total direct expenses in 2011, 2012, and 2013.
- Allocated expenses are truly allocated. We have chosen to make them 28% of your direct expenses.
- In all four years, subtract your direct and allocated expenses from your gross margin to get your net income. Calculate the return on sales (ROS = (Net income)/Revenue ) for all four years.
- Finally, add up the net income across all four years to get your total net income for the Exercise. This is your final, quantitative, bottom-line result.

## COMPARISON OF RESULTS

If you are in a classroom setting, naturally the teams will compare how they did relative to each other. But, in an absolute sense, what is a "good" result? Since results from this exercise are on a continuum, there is not a meaningful threshold that separates "good" from "bad." But Figure F.1 shows the distribution of results from eighteen different teams that did this Exercise in classroom settings.

## EXPLAINING YOUR RESULTS

There are many factors that contribute to your results. If you go back to the first module, your strategy accounts for some of what you achieved.

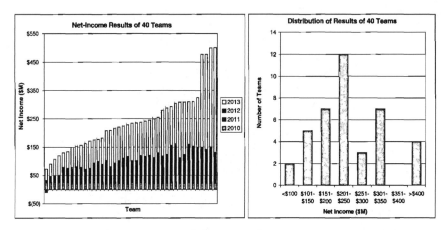

**Figure F.1** Results from some other teams who have done this exercise.

For example, you may have chosen to attack the Consumer market. What happens in the Exercise is that the Consumer market develops fairly rapidly, but not as rapidly as any of the consultants have projected. Therefore the addressable market was smaller than planned, so revenues will be less than expected. (This often happens in the marketplace—the potential consumer market is frequently over-estimated in the early years.)

Another item that significantly affects results is how well the product is matched to one or more of the market segments. You may have had difficulty in reaching your cost goals, and so compromised on some of the "Basic" features that you put into your product. And you might have noticed that it is very difficult to come up with a hybrid product that is well suited to two, or all three, of the market segments.

Yet another factor that affects your results is the price you charged. Each of the market segments has different degrees of price sensitivity, and each expects a different rate of price decline from year to year. Nevertheless, while price is an important factor in the buying decision, in this simulation it is only one factor among seven others.

Of course, your costs and gross margins affect your results significantly. You may have been unable to reach your cost targets, and so decided to live with a smaller margin. This goes directly to your bottom line, and may explain some differences among the teams.

Finally, your other expenses do have an impact on your results. You may have chosen a design with low-cost parts, but the annual costs of using those parts may have been relatively high. Nevertheless, in this exercise these direct expenses are of a magnitude such that they constitute at most a second-order effect.

# Get Financial Results

**AFTER COMPLETING MODULE F**

If you are in a classroom setting we strongly encourage the teams to share their final results with each other. They also should try to account for any significant differences in results among the teams. Either as an individual reader or as a member of a class, depending on what information you obtained, here are some of the things you should have learned by the end of Module F:

In Module A
- Your business scenario.
- *Your company's* product and name.
- Your business strategy.
- Last year's financial results.
- Market forecasts—sizes, shares.
- Competitive strengths and weaknesses.
- Competitors' costs.

In Module B
- Customer preferences.
- Market-feature table.
- Conjoint Analysis: buyer utilities.
- Product attributes: importance-value table.
- Features that will be in your product.

In Module C
- Historical and current prices for your current product.
- This year's projected financial results.
- Price trends.
- Price experience curve.
- Your planned gross margins.
- Cost experience curve.
- Target prices and costs (COGS) for 2011–2013.

In Module D
- Function–subsystem matrix.
- Current costs of subsystems and components.
- Cost targets for each subsystem.
- Subsystem cost gaps (2011–2013).

In Module E
- Final attributes or features of your new product.
- Final selection of components and suppliers.
- Volume and partnership discounts.
- COGS for new product (2011–2013).
- Prices for new product (2011–2013).
- Gross margins for new product (2011–2013).
- Development expenses (2010).
- Yearly expenses (2011–2013).

In Module F
- Market share (2010–2013).
- Unit sales and revenues (2010–2013).
- Net income (2010–2013).
- How you did, relative to competitors.
- Reasons for your results.

> You have completed Module F.
> Please continue to Module G.

# Module G

Optimizing Results

In the Exercise you considered the available data and made decisions to optimize your company's performance. Most likely, you purchased enough information to be able to achieve a good result. However, you may have been confused by extraneous information or missing information needed to make informed decisions. Limited and extraneous data are typical in real-life problems because gathering complete and meaningful data is either difficult or expensive. In the Exercise, you competed against three competitors' products. Although this is typical of real problems, you rarely know as much about the competition's product as assumed for the simulator.

In this module, we extend the results you achieved in the Exercise and show a method to optimize the results. Building off the model presented here, you can apply this method to solve other problems you encounter. By changing the inputs to your attributes and levels, you can use this structure and method to gain insights about and solve your problem.

**THE CHALLENGE**

Usually your challenge is to design the right product at the right cost to optimize your company's performance *without* detailed knowledge of competitors' *planned* products. With the methods described in this module, you can model key characteristics of the marketplace and of your product design to solve for the optimum design, thereby saving significant time and expense. Besides providing a solution without having the difficult-to-come-by competitive data, the approach described here makes it possible to obtain solutions quickly by using a computer to solve the optimization problem. Although simple to visualize (Fig. G.1), the analysis may appear complicated since it involves a spreadsheet of several matrices and uses an add-on Microsoft Excel® function called Solver.

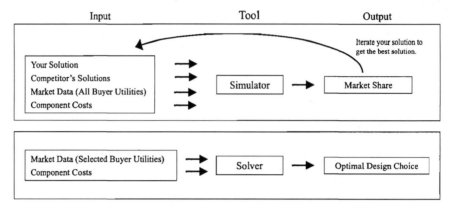

**Figure G.1** Differences between market simulation approach of exercise in Modules A through F and the solver approach described here in Module G.

Learning to use this spreadsheet is a bit like tying shoelaces—it is easier to copy than to explain. Step by step, we will present the integer programming method used to solve for the design that achieves either maximum market share, maximum profit, or meets the target cost, depending on your objective. This approach organizes key inputs to focus your resources on the issues most relevant to your desired results. The benefits are:

- A framework for making design tradeoffs.
- Likely solutions to save R&D time and expense.
- A sensitivity analysis for showing the sensitivity of the optimal solutions to the input variables.

## INPUTS

The main inputs are the Conjoint Analysis buyer utilities that describe the market needs and the costs of the components available to build your product. Typically you know the key attributes that a customer considers when making the purchase decision for your product and what subsystems you might need to provide that functionality. We use the attribute mapping matrix to map the desirable customer features (attributes) to the subsystems making up the product. With a reasonable effort you will also be able to determine the component costs that make up the subsystems. From Conjoint Analysis you will have the buyer utilities, describing the marketplace.

## OUTPUTS

The solver determines the design and price choices that:

- optimize market share, or
- optimize profit, or
- meet the target cost.

# Optimizing Results

We will start with an initial guess. Once you set up the problem in the manner we describe, the computer solves for the optimum solution using an integer programming iteration method. Most spreadsheets have this capability in an add-in function. We have found Microsoft Excel's add-on Solver convenient.

## SOLUTION

Profit before taxes is defined as product revenue less product costs and other expenses. Mathematically:

$$\text{Profit } P = [(p - c)v] - e \tag{1}$$

where $p$ is unit price, $c$ is unit cost, $v$ is volume or quantity sold, and $e$ is corresponding total product expense. Once a design is defined, it is relatively easy to calculate its final price, cost, volume, and expenses. To determine the design, we will consider a wide range of possibilities and then determine the optimum solution. In other words, we will start with an initial guess and then find the optimum solution. To express the wide range of possibilities, we will use matrices to organize the information.

The objective is to maximize profit, $P$. To do so, we will develop the terms of this profit expression in a logical framework that can be adapted to other problems. Each term on the right-hand side of the equation will be expanded into a matrix format. These matrices line up with the product attributes and levels of the Conjoint Analysis survey. To illustrate this concept, we will use the wrist videophone example, simplifying to the first year and for the government market only. The key matrices are shown in Fig. G.2. The model used to solve for the design that maximizes profit is a spreadsheet of these matrices. We will develop this in general terms so that you can apply this approach to your specific cases. The inputs come from Conjoint Analysis survey outputs and detailed cost data for components used to make the product.

## SETTING UP THE PROBLEM

Let us look at the wrist videophone example as we consider each term, step by step. Start by creating a solution framework using the Conjoint Analysis attributes and levels of your customer survey. In a spreadsheet, list attributes across the top and levels down the side. Include the complete description as presented to the customer (e.g., custom color for $25 more), as shown in Fig. G.3.

Below this, list the other matrices. They are defined as follows:

$P =$ table of prices expressed as base price and incremental prices for each attribute level as defined in the Conjoint Analysis survey.
$\mathbf{U} =$ buyer utility matrix with outputs from the Conjoint Analysis.

**Figure G.2** Spreadsheets of key matrices used to solve for optimum wrist videophone design in first year based on government segment data only.

Product Description Table

|  | Attribute 1 | Attribute 2 | Attribute 3 | Attribute 4 | Attribute 5 | Attribute 6 | Attribute 7 | Attribute 8 |
|---|---|---|---|---|---|---|---|---|
|  | Color | Display Size | Video Speed | Stored Numbers | Talk Time | Messaging | Delivery | Base Price |
| Level 1 | Chrome for $5 discount | 1.5" B&W | 2 frames/s for $25 discount | 4 numbers | 15 minutes | none | 24 hours for $15 | $ 350 |
| Level 2 | Black | 1.5" color for $25 | 12 frames/s | 16 numbers for $10 | 30 minutes for $40 | 2 minutes for $25 | 72 hours | $ 400 |
| Level 3 | Custom for $25 | 2.0" color for $50 | 24 frames/s for $50 | 32 numbers for $25 | 60 minutes for $75 | 5 minutes for $40 |  | $ 450 |
| Level 4 | Gold for $35 |  |  |  |  |  |  | $ 500 |
| Level 5 |  |  |  |  |  |  |  | $ 550 |

**Figure G.3** Wrist videophone product description table showing all product possibilities presented to customers.

## Optimizing Results

$\mathbf{X}$ = design and price choices ("X" because unknown in the beginning).

The other two are a combination of a mapping matrix and either a cost or an expense matrix.

$\mathbf{S}$ = subsystem mapping matrix. It is a variation of the function–subsystem mapping matrix. It provides the fraction of each attribute allocated to each subsystem.

$\mathbf{C}$ = attribute cost matrix. It contains best subsystem costs based on the subsystem mapping matrix, S.

$\mathbf{E}$ = attribute expense matrix. It contains the corresponding expenses for each subsystem based on the subsystem mapping matrix, S.

---

**For Practitioners: Terminology Used in the Matrices.** We will use the following convention. A matrix $\mathbf{A}$ of $m$ rows and $n$ columns will be represented as $\mathbf{A}^{[m,\,n]}$ or just plain $\mathbf{A}$ when the size is clear. The index superscript is used to prevent confusion when the last column is not used in some of the matrices. The indices $k$, $l$, $m$, and $n$ will denote the size. The indices $i$ and $j$ will denote the location. The terms of the matrix are $a_{ij}$ with row index $i$ ranging from one to $m$ and the column index $j$ ranging from 1 to $n$, as follows:

$$\mathbf{A}^{[m,\,n]} = \begin{bmatrix} a_{11} & a_{12} & \cdots & a_{1n} \\ a_{21} & a_{22} & \cdots & a_{2n} \\ & & \cdots & \\ a_{m1} & a_{m2} & \cdots & a_{mn} \end{bmatrix}$$

We will use two matrix operations, matrix multiplication and inner product. In matrix multiplication, the sum-product is used to combine terms. For example, when

$$\mathbf{C}^{[k,\,l]} = \mathbf{A}^{[k,\,m]} \mathbf{B}^{[m,\,l]},$$

the individual terms are

$$c_{ij} = a_{i1}b_{1j} + a_{i2}b_{2j} + \cdots + a_{im}b_{mj}.$$

The inner product of two matrices yields a scalar. For an inner product the corresponding terms in two like-dimensioned matrices are multiplied together and then all terms summed as follows:

$$c = \mathbf{A}^{[m,\,n]} \circ \mathbf{B}^{[m,\,n]} = a_{11}b_{11} + a_{12}b_{12} + \cdots + a_{mn}b_{mn}$$

**Attribute Price Differences by Level, P** $^{[k, l]}$

|  | Attribute 1 Color | Attribute 2 Display Size | Attribute 3 Video Speed | Attribute 4 Stored Numbers | Attribute 5 Talk Time | Attribute 6 Messaging | Attribute 7 Delivery | Attribute 8 Base Price |
|---|---|---|---|---|---|---|---|---|
| Level 1 | $ (5) | $ - | $ (25) | $ - | $ - | $ - | $ 15 | $ 350 |
| Level 2 | $ - | $ 25 | $ - | $ 10 | $ 40 | $ 25 | $ - | $ 400 |
| Level 3 | $ 25 | $ 50 | $ 50 | $ 25 | $ 75 | $ 40 | $ 1,000 | $ 450 |
| Level 4 | $ 35 | $ 1,000 | $ 1,000 | $ 1,000 | $ 1,000 | $ 1,000 | $ 1,000 | $ 500 |
| Level 5 | $ 1,000 | $ 1,000 | $ 1,000 | $ 1,000 | $ 1,000 | $ 1,000 | $ 1,000 | $ 550 |

**Figure G.4** Base price and incremental differences for each level included in the Conjoint Analysis survey. Undefined terms assigned $1000.

**Buyer Utilities Matrix, U** $^{[k, l]}$

|  | Attribute 1 Color | Attribute 2 Display Size | Attribute 3 Video Speed | Attribute 4 Stored Numbers | Attribute 5 Talk Time | Attribute 6 Messaging | Attribute 7 Delivery | Attribute 8 Base Price |
|---|---|---|---|---|---|---|---|---|
| Level 1 | 8 | 9 | 0 | 0 | 0 | 33 | 0 | 65 |
| Level 2 | 71 | 69 | 42 | 45 | 70 | 17 | 23 | 72 |
| Level 3 | 12 | 42 | 46 | 15 | 65 | 15 | 0 | 48 |
| Level 4 | 1 | 0 | 0 | 0 | 0 | 0 | 0 | 32 |
| Level 5 | 0 | 0 | 0 | 0 | 0 | 0 | 0 | 1 |
| max | 71 | 69 | 46 | 45 | 70 | 33 | 23 | 72 |

**Figure G.5** Wrist videophone buyer utilities for government market segment.

Below the product description table, put the price matrix, $P^{[k,l]}$ (Fig. G.4). The index $l$ is the number of attributes and $k$ is the maximum number of levels. For convenience, we keep the base prices in the last column, $l$. This will be useful later when we want to separate the cost of various functions from the base price. So $P^{[k, l]}$ contains the base price in the last column and the incremental prices for the other attributes at all levels directly from the Conjoint Analysis survey at any place where a level is not defined in the price matrix, include a very high value so all terms have a value for the solver, but the high value is never selected for a solution. For illustrative purposes, $1000 was used for undefined levels since this is much greater than the base price.

Next, set up the buyer utilities matrix, **U**. This particular one (Fig. G.5) is for the government market segment results only.

Next, we have the design choices matrix, $X^{[k, l]}$, that you need to solve for. (*Note*: In this text, $X^{[k, m]}$ is the same matrix with the last column (base price), $l$, omitted. This will be useful later.) The values in this matrix are the unknowns. If a level is selected, a 1 appears in the matrix. All other levels must be 0 for each attribute so the vertical sums are 1. There must be an initial guess for each attribute. For illustrative purposes, all level 1's are shown as a starting point (Fig. G.6).

At this point, we are ready to build the product unit price term in the model from the product choice and price matrix as follows:

$$\text{Unit price } p = \mathbf{X} \circ \mathbf{P} \tag{2}$$

# Optimizing Results

**Product Choices, $X^{[k,l]}$**

|         | Attribute 1 Color | Attribute 2 Display Size | Attribute 3 Video Speed | Attribute 4 Stored Numbers | Attribute 5 Talk Time | Attribute 6 Messaging | Attribute 7 Delivery | Attribute 8 Base Price |
|---------|---|---|---|---|---|---|---|---|
| Level 1 | 1 | 1 | 1 | 1 | 1 | 1 | 1 | 1 |
| Level 2 | 0 | 0 | 0 | 0 | 0 | 0 | 0 | 0 |
| Level 3 | 0 | 0 | 0 | 0 | 0 | 0 | 0 | 0 |
| Level 4 | 0 | 0 | 0 | 0 | 0 | 0 | 0 | 0 |
| Level 5 | 0 | 0 | 0 | 0 | 0 | 0 | 0 | 0 |
| sum     | 1 | 1 | 1 | 1 | 1 | 1 | 1 | 1 |

**Figure G.6** Wrist videophone product choice matrix with initial values all set to level 1.

**Prices per Unit by Attribute ($).** $p = X^{[k,l]} \circ P^{[k,l]}$

|         | Attribute 1 Color | Attribute 2 Display Size | Attribute 3 Video Speed | Attribute 4 Stored Numbers | Attribute 5 Talk Time | Attribute 6 Messaging | Attribute 7 Delivery | Attribute 8 Base Price |  |
|---|---|---|---|---|---|---|---|---|---|
| Level 1 | $ (5) | $ - | $ (25) | $ - | $ - | $ - | $ 15 | $ 350 | |
| Level 2 | $ - | $ - | $ - | $ - | $ - | $ - | $ - | $ - | |
| Level 3 | $ - | $ - | $ - | $ - | $ - | $ - | $ - | $ - | |
| Level 4 | $ - | $ - | $ - | $ - | $ - | $ - | $ - | $ - | |
| Level 5 | $ - | $ - | $ - | $ - | $ - | $ - | $ - | $ - | Unit Price, p |
| sum     | $ (5) | $ - | $ (25) | $ - | $ - | $ - | $ 15 | $ 350 | $ 335 |

**Figure G.7** Calculation of unit price from matrices for a wrist videophone design with all level 1 values in choice matrix set to 1.

As stated in Eq. (2), the product of the design choice and price matrix defines the unit price, $p$, of the selected design. In general, this is represented as $p = X \circ P$ and is depicted in Fig. G.7. For the government data example, where level 1 values were selected for all attributes, the unit price $p = (-\$5 + 0 - \$25 + 0 + 0 + 0 + \$15 + \$350) = \$335$.

Now, build up the terms for unit cost and total expense matrices. These terms are more complicated because they combine several matrices. The terms of the profit equation are defined as follows:

$$\text{Unit cost } c = X \circ (C\,S), \quad \text{and similarly} \qquad (3)$$

$$\text{Total expense } e = X \circ (E\,S) \qquad (4)$$

The product cost terms come from available component costs and the subsystem mapping matrix for each design possibility at the attribute level. *This is the most critical step in the profit optimization process.* It is here that you select, and then begin to narrow down the costs from all the possibilities. You will need to create an attribute cost matrix, $C^{[k,m]}$, for all feasible design possibilities. To create the attribute cost matrix, you will need both the best subsystem costs for each level in the product description table and a variation of the function–subsystem matrix we derive here. Start with the function–subsystem mapping matrix, $F^{[n,m]}$ (Fig. G.8, from Module D's Item A). Recall that the attribute percentages were developed with the design team to map the subsystems to the customer needs (see Chapter 4).

**Function/Subsystem Mapping Matrix, F$^{[n,m]}$**

| Attribute / Subsystem | Attribute 1 Color | Attribute 2 Display Size | Attribute 3 Video Speed | Attribute 4 Stored Numbers | Attribute 5 Talk Time | Attribute 6 Messaging | Attribute 7 Delivery | row sum, t |
|---|---|---|---|---|---|---|---|---|
| Display | 60% | 65% | 25% | 0% | 10% | 0% | 10% | 170% |
| Controller | 0% | 0% | 50% | 50% | 30% | 40% | 10% | 180% |
| Camera | 0% | 15% | 15% | 0% | 0% | 0% | 10% | 40% |
| Telephone | 0% | 0% | 10% | 20% | 30% | 40% | 10% | 110% |
| Power | 0% | 20% | 0% | 30% | 20% | 15% | 10% | 95% |
| Housing & Assembly | 40% | 0% | 0% | 0% | 10% | 5% | 50% | 105% |
| column sum | 100% | 100% | 100% | 100% | 100% | 100% | 100% | |

**Figure G.8** Wrist videophone function–subsystem matrix plus row sum and column sum.

**Subsystem Mapping Matrix, S$^{[n,m]}$**

| Attribute / Subsystem | Attribute 1 Color | Attribute 2 Display Size | Attribute 3 Video Speed | Attribute 4 Stored Numbers | Attribute 5 Talk Time | Attribute 6 Messaging | Attribute 7 Delivery | check sum |
|---|---|---|---|---|---|---|---|---|
| Display | 35% | 38% | 15% | 0% | 6% | 0% | 6% | 100% |
| Controller | 0% | 0% | 28% | 28% | 17% | 22% | 6% | 100% |
| Camera | 0% | 38% | 38% | 0% | 0% | 0% | 25% | 100% |
| Telephone | 0% | 0% | 9% | 18% | 27% | 36% | 9% | 100% |
| Power | 0% | 21% | 0% | 32% | 21% | 16% | 11% | 100% |
| Housing & Assembly | 38% | 0% | 0% | 0% | 10% | 5% | 48% | 100% |

**Figure G.9** Subsystem mapping matrix for the wrist videophone exercise.

Now sum the rows in the function–subsystem matrix to get a subsystem total, called $t$:

$$t_i = \sum_{j=1,s} f_{ij} \tag{5}$$

The subsystem totals are used to calculate a variation of the function–subsystem matrix where the terms are expressed as a percent of the subsystem rather than the attribute. For each term in the new mapping matrix, $S^{[n,m]}$, calculate the portion of the subsystem (row) total allocated to each attribute (column) as follows:

$$s_{ij} = a_{ij}/t_i \tag{6}$$

For the government example this calculation yields the percents in the subsystem mapping matrix S in Fig. G.9.

Now combine these ratios with the component cost data. Individual component costs sum to subsystem costs. You will need to collect cost data for each design possibility. Usually the design or purchasing members of your team will be able to provide this data. You will need to collect cost for every possible solution. A simple way to do this is to collect costs for each row in the product description table. When summarized, it looks like the quotations provided in Module E. Select data to meet

## Optimizing Results

| Subsystem (or part) / Criteria Satisfied | Unit Cost for |  |  |  |
|---|---|---|---|---|
|  | level 1 | level 2 | level 3 | level 4 |
| Display | $ 27.25 | $ 50.75 | $ 75.50 | $ 77.00 |
| bezel | $ 3.25 | $ 4.75 | $ 6.50 | $ 8.00 |
| LCD | $ 8.00 | $ 28.00 | $ 49.00 | $ 49.00 |
| driver circuit | $ 16.00 | $ 18.00 | $ 20.00 | $ 20.00 |
| Controller | $ 41.00 | $ 44.75 | $ 58.75 | $ 41.00 |
| Camera | $ 28.00 | $ 34.00 | $ 33.00 | $ 28.00 |
| Telephone | $ 36.00 | $ 36.00 | $ 36.00 | $ 36.00 |
| Power | $ 16.00 | $ 20.00 | $ 16.00 | $ 16.00 |
| Housing & Assembly | $ 13.00 | $ 14.00 | $ 13.00 | $ 13.00 |
| Total | $ 161.25 | $ 199.50 | $ 232.25 | $ 211.00 |

**Figure G.10** Wrist videophone cost data that meet the criteria of the levels in the Conjoint Analysis survey. (Parts used for Display are listed separately, as could be done for each subsystem.)

the requirements for each row in the product description matrix. Fig. G.10 shows representative cost data for wrist videophones suited to the government market.

Here you must apply your other insights about the components being considered to meet the product requirements. We assume that choices for one attribute are independent of the others. That is, if you select level 1 for color (black), it does not affect another attribute's other levels (e.g. 2 in. display). However, there usually *are* direct correlations at the component level. For example, in the wrist videophone exercise, only the 2012 processor works with the 12 fps ASICs to provide 12 frames/second video speeds. In such cases, make sure that the components selected meet the requirements for the product choice. When there are several components that would meet the criteria, choose the lowest-cost one, as was done for the telephone subsystem in this example.

This table of cost data becomes the best subsystem cost matrix, **B**. Choose the subsystem totals and orient the matrix so that all the cost data for each level is in a row. You will have data for at least the first two rows of the matrix. However, since you may not have more than two levels for some of the attributes, you will not have meaningful data for all rows. At this stage, we need a value to carry out the matrix multiplication. Use the best costs from previous rows. The rationale is based on the fact that all subsystems are required to build up a complete product, but the undefined terms should not contribute more than the best cost for the undefined subsystems. The undefined terms will be replaced later. As an illustration, Fig. G.11 shows a summary of the best cost data from the wrist videophone exercise for component choices that support all of one level. Level 3 data for housing and assembly is assumed equivalent to the best cost of level 1. Similarly, terms are filled in for the missing sections for level 4. Since level 5 is only used for price, it is left off. We will deal with this momentarily.

**Best Subsystem Costs for Each Level, B$^{[k,n]}$**

|  | Display | Controller | Camera | Telephone | Power | Housing & Assembly |
|---|---|---|---|---|---|---|
| Level 1 | $ 27.25 | $ 41.00 | $ 28.00 | $ 36.00 | $ 16.00 | $ 13.00 |
| Level 2 | $ 50.75 | $ 44.75 | $ 34.00 | $ 36.00 | $ 20.00 | $ 14.00 |
| Level 3 | $ 75.50 | $ 58.75 | $ 33.00 | $ 36.00 | $ 16.00 | $ 13.00 |
| Level 4 | $ 77.00 | $ 41.00 | $ 28.00 | $ 36.00 | $ 16.00 | $ 13.00 |
| Level 5 |  |  |  |  |  |  |

**Figure G.11** Best subsystem costs from component roll-ups based on each level's wrist videophone description.

**Attribute Cost Matrix, C$^{[k,n]}$ = B$^{[k,n]}$ S$^{[n,m]}$**

| Attribute / Level | Attribute 1 Color | Attribute 2 Display Size | Attribute 3 Video Speed | Attribute 4 Stored Numbers | Attribute 5 Talk Time | Attribute 6 Messaging | Attribute 7 Delivery | Attribute 8 Base Price |
|---|---|---|---|---|---|---|---|---|
| Level 1 | $ 15 | $ 24 | $ 29 | $ 23 | $ 23 | $ 25 | $ 22 |  |
| Level 2 | $ 23 | $ 36 | $ 36 | $ 25 | $ 26 | $ 27 | $ 26 |  |
| Level 3 | $ 32 | $ 45 | $ 43 | $ 28 | $ 29 | $ 29 | $ 1,000 |  |
| Level 4 | $ 32 | $ 1,000 | $ 1,000 | $ 1,000 | $ 1,000 | $ 1,000 | $ 1,000 |  |
| Level 5 | $ 1,000 | $ 1,000 | $ 1,000 | $ 1,000 | $ 1,000 | $ 1,000 | $ 1,000 |  |

**Figure G.12** Attribute cost matrix for wrist videophone that satisfies government market segment preferences.

Now you are ready to calculate the costs. Multiply the subsystem mapping by the best subsystem costs matrix to create an attribute cost matrix as follows:

$$\mathbf{C}^{[k,m]} = \mathbf{B}^{[k,n]} \mathbf{S}^{[n,m]} \qquad (7)$$

After multiplying, replace the undefined terms with a high value so the solver does not select them. For the government market segment in the wrist videophone example, this matrix multiplication gives the attribute cost matrix (Fig. G.12).

The inner product of the design choice and the attribute cost matrix is the unit cost of the selected design,

$$\text{Total cost } c = \left\{ \mathbf{X}^{[k,m]} \circ \mathbf{C}^{[k,m]} \right\} \qquad (8)$$

(Note here that $m = l - 1$ so the price column of $\mathbf{X}^{[k,l]}$ is not used.) Assuming all level 1 choices for the government market segment of the wrist videophone exercise, we get $161 (Fig. G.13).

Similarly, use the corresponding expenses for the components selected to create an expense matrix. Then multiply the subsystem mapping matrix by the best expense matrix and take the inner product

## Optimizing Results

**Production Costs Per Unit ($),** $c = \{X^{[k,m]} \circ C^{[k,m]}\}$

|  | Attribute 1 Color | Attribute 2 Display Size | Attribute 3 Video Speed | Attribute 4 Stored Numbers | Attribute 5 Talk Time | Attribute 6 Messaging | Attribute 7 Delivery | Attribute 8 Base Price |  |
|---|---|---|---|---|---|---|---|---|---|
| Level 1 | $ 15 | $ 24 | $ 29 | $ 23 | $ 23 | $ 25 | $ 22 |  |  |
| Level 2 | $ - | $ - | $ - | $ - | $ - | $ - | $ - |  |  |
| Level 3 | $ - | $ - | $ - | $ - | $ - | $ - | $ - |  |  |
| Level 4 | $ - | $ - | $ - | $ - | $ - | $ - | $ - |  |  |
| Level 5 | $ - | $ - | $ - | $ - | $ - | $ - | $ - |  | Total Unit Cost, c |
| sum | $ 15 | $ 24 | $ 29 | $ 23 | $ 23 | $ 25 | $ 22 |  | $ 161 |

**Figure G.13** Total unit cost calculation for all level 1 choices for the government market segment in the wrist videophone exercise.

**Expense Costs for Unit ($M),** $e = X^{[k,m]} \circ (E^{[k,n]} S^{[n,m]})$

|  | Attribute 1 Color | Attribute 2 Display Size | Attribute 3 Video Speed | Attribute 4 Stored Numbers | Attribute 5 Talk Time | Attribute 6 Messaging | Attribute 7 Delivery | Attribute 8 Base Price |  |
|---|---|---|---|---|---|---|---|---|---|
| Level 1 | $ 1 | $ 1 | $ 2 | $ 2 | $ 2 | $ 2 | $ 2 |  |  |
| Level 2 | $ - | $ - | $ - | $ - | $ - | $ - | $ - |  |  |
| Level 3 | $ - | $ - | $ - | $ - | $ - | $ - | $ - |  |  |
| Level 4 | $ - | $ - | $ - | $ - | $ - | $ - | $ - |  |  |
| Level 5 | $ - | $ - | $ - | $ - | $ - | $ - | $ - |  | Total Expense, e |
| sum | $ 1 | $ 1 | $ 2 | $ 2 | $ 2 | $ 2 | $ 2 |  | $ 12 |

**Figure G.14** Total expense calculation for all level 1 choices for the government market segment in the wrist videophone exercise.

with the design matrix to calculate the total expense as follows:

$$\text{Total expense } e = \mathbf{X}^{[k,m]} \circ (\mathbf{E}^{[k,n]} \mathbf{S}^{[n,m]}) \tag{9}$$

where $\mathbf{E}^{[k,n]}$ is the subsystem level expense matrix analogous to the subsystem level unit cost matrix. So for all level 1 choices of the government market segment of the wrist videophone exercise, we get $12million (Fig. G.14).

We now have all of the terms in the profit expression except volume. You will need an estimate of the forecasted total addressable market. There will be competitors with similar products, but this model assumes no knowledge of their products. In this approach, we introduce an efficiency term based on the buyer utilities to model the effectiveness of your design in the marketplace. This will not yield a market share, but it will allow relative comparisons for a given forecast. So, volume is a function of total addressable market and design efficiency. We model volume, $v$, as the product of market efficiency, $\eta$, and a forecast, $f$, as follows:

$$v = \eta f \tag{10}$$

Although this looks simple, this term makes the expression for profit nonlinear because it is a function of the unknown, $\mathbf{X}$. The efficiency, $\eta$, is determined using the buyer utilities and is a function of the design choice. Efficiency, $\eta$, is determined from the total buyer utility for the design

**Buyer Utilities for Selection Produced & Sold,** $u = X^{[k,l]} \circ U^{[k,l]}$

|  | Attribute 1 | Attribute 2 | Attribute 3 | Attribute 4 | Attribute 5 | Attribute 6 | Attribute 7 | Attribute 8 |  |
|---|---|---|---|---|---|---|---|---|---|
|  | Color | Display Size | Video Speed | Stored Numbers | Talk Time | Messaging | Delivery | Base Price |  |
| Level 1 | 8 | 9 | 0 | 0 | 0 | 33 | 0 | 65 |  |
| Level 2 | 0 | 0 | 0 | 0 | 0 | 0 | 0 | 0 |  |
| Level 3 | 0 | 0 | 0 | 0 | 0 | 0 | 0 | 0 |  |
| Level 4 | 0 | 0 | 0 | 0 | 0 | 0 | 0 | 0 |  |
| Level 5 | 0 | 0 | 0 | 0 | 0 | 0 | 0 | 0 | total utility, u |
| sum | 8 | 9 | 0 | 0 | 0 | 33 | 0 | 65 | 115 |

**Figure G.15** Sum of buyer utilities for a government market solution to wrist videophone exercise.

choice selected divided by the maximum total buyer utility as follows:

$$\text{Total utility } u = \mathbf{X}^{[k,l]} \circ \mathbf{U}^{[k,l]} \quad (11)$$

$$\text{Design efficiency } \eta = u \bigg/ \sum_{j=1,r} u_{j|\max} \quad (12)$$

where $u_{j|\max}$ is the maximum buyer utility for the attribute, $j$, in the utility matrix of Fig. G.5. Here (Fig. G.15) we sum over all utilities, including price. So, using all 1's in the government market of the wrist videophone exercise, the efficiency $\eta = 115/429 = 27\%$.

Therefore, the expanded profit expression of Eq. (1) in matrix form is

$$\text{Profit } P = [(p-c)v] - e$$
$$= \left[ (\{\mathbf{X} \circ \mathbf{P}\} - \{\mathbf{X} \circ (\mathbf{C}\,\mathbf{S})\})(\{\mathbf{X} \circ \mathbf{U}\} \bigg/ \sum_{j=1,r} u_{j|\max})f \right] - \{\mathbf{X} \circ (\mathbf{E}\,\mathbf{S})\} \quad (13)$$

or, showing the size of the matrices,

$$= \left[ (\{\mathbf{X}^{[k,l]} \circ \mathbf{P}^{[k,l]}\} - \{\mathbf{X}^{[k,m]} \circ (\mathbf{C}^{[k,n]} \mathbf{S}^{[n,m]})\})(\{\mathbf{X}^{[k,l]} \circ \mathbf{U}^{[k,l]}\} \bigg/ \sum_{j=1,r} u_{j|\max})f \right]$$
$$- \{\mathbf{X}^{[k,m]} \circ (\mathbf{E}^{[k,n]} \circ \mathbf{S}^{[n,m]})\}$$

We will use this objective function to solve for the design that maximizes profit. Note that since $\eta$ is a function of $\mathbf{X}$, this objective function is nonlinear.

## CONSTRAINTS

We now have all the terms of the profit equation. Now add the following constraints:

- All design choices must be positive integers:

$$x_{ij} \geq 0 \quad (14)$$
$$x_{ij} = \text{integer} \quad (15)$$

## Optimizing Results

- The sum of the design choices for an attribute must equal 1 (i.e., only one choice permitted for each attribute):

$$\sum_{i=1,q} x_{ij} = 1, \text{for each attribute } r \tag{16}$$

Use an iterative solver to solve for the design, **X**, that maximizes the profit and satisfies the constraints. Refer to the manual that comes with your spreadsheet solver for how to use this capability. You will need to specify the cell for total profit and tell the solver to maximize profit by changing the design matrix, **X**, subject to the constraints.

## RESULTS AND OBSERVATIONS

The essential solver spreadsheets for a problem of the wrist videophone exercise's complexity fit on one page, as shown in Fig. G.16. For illustrative purposes, this spreadsheet covers only the first year for the government market. Once you have created this spreadsheet, tell the solver to find the terms of the design choice matrix, **X**, such that total profit is optimized. The solver routine begins with the initial guess and uses an optimization algorithm that optimizes the value of the objective function while ensuring that the constraints remain satisfied. Running the solver routine usually takes less than a minute as it quickly eliminates many possibilities that would require significant investment to understand thoroughly.

In this case, the result for maximum profit has the following characteristics:

| Attribute | Level | Price |
|---|---|---|
| Color | Black | |
| Display size | 1.5 in. color | $25 |
| Video speed | 24 frames/second | 50 |
| Stored numbers | 16 numbers | 10 |
| Talk time | 60 minutes | 75 |
| Messaging | 5 minutes | 40 |
| Delivery | 72 hours | |
| Base price: | | 550, |
| Total price: | | $750 |

This solution is depicted by 1's in the appropriate cells of the product choices matrix, **X**, as shown in Fig. G.16. The price is determined by adding the appropriate premiums to the base price using the attribute price difference matrix, **P**. For the wrist videophone example under consideration, the cost is $212.

**Figure G.16** Essential solver spreadsheets for one year of wrist videophone using government segment data.

This solution is the absolute maximum for this scenario, but it is hard to know for sure by just running the solver once. That is because the solution found can depend on the starting point you define. Depending on the shape of the surface in the vicinity of the starting point, it is possible that the solver will find a local maximum rather than the absolute maximum. For this reason it is important to perturb the starting point and run the design choice solver several times. We recommend starting with all 1's in for the first level and also trying all 1's in the highest level. In most cases, you will quickly solve for the design that yields optimum profit.

To illustrate the range of solutions, we will use a three-dimensional graph that depicts profit as a surface. Fig. G.18 is a contour plot of profit as a function of price and cost for the wrist videophone. The data for this

# Optimizing Results

**Calculated** — Production Costs per Unit ($). $c = \{X^{[k,m]} \circ C^{[k,m]}\}$

|  | Attribute 1 Color | Attribute 2 Display Size | Attribute 3 Video Speed | Attribute 4 Stored Numbers | Attribute 5 Talk Time | Attribute 6 Messaging | Attribute 7 Delivery | Attribute 8 Base Price |  |
|---|---|---|---|---|---|---|---|---|---|
| Level 1 | $ - | $ - | $ - | $ - | $ - | $ - | $ - |  |  |
| Level 2 | $ 23 | $ 36 | $ - | $ 25 | $ - | $ - | $ 26 |  |  |
| Level 3 | $ - | $ - | $ 43 | $ - | $ 29 | $ 29 | $ - |  |  |
| Level 4 | $ - | $ - | $ - | $ 0 | $ - | $ 0 | $ 0 |  |  |
| Level 5 | $ - | $ - | $ 0 | $ - | $ - | $ - | $ 0 |  | unit cost, c |
| sum | $ 23 | $ 36 | $ 43 | $ 25 | $ 29 | $ 29 | $ 26 |  | 211.9375 |

**Calculated** — Expense Costs for Unit ($M). $e = X^{[k,m]} \circ (E^{[k,n]} S^{[n,m]})$

|  | Attribute 1 | Attribute 2 | Attribute 3 | Attribute 4 | Attribute 5 | Attribute 6 | Attribute 7 | Attribute 8 |  |
|---|---|---|---|---|---|---|---|---|---|
| Level 1 | $ - | $ - | $ - | $ - | $ - | $ - | $ - |  |  |
| Level 2 | $ 3 | $ 3 | $ - | $ 2 | $ - | $ - | $ 3 |  |  |
| Level 3 | $ - | $ - | $ 3 | $ - | $ 2 | $ 2 | $ - |  |  |
| Level 4 | $ - | $ - | $ - | $ 0 | $ - | $ 0 | $ 0 |  |  |
| Level 5 | $ - | $ - | $ 0 | $ - | $ - | $ - | $ 0 |  | Total Expense, e |
| sum | $ 3 | $ 3 | $ 3 | $ 2 | $ 2 | $ 2 | $ 3 |  | 16.80816 |

**Calculated** — Prices per Unit by Attribute ($). $p = X^{[k,l]} \circ P^{[k,l]}$

|  | Attribute 1 | Attribute 2 | Attribute 3 | Attribute 4 | Attribute 5 | Attribute 6 | Attribute 7 | Attribute 8 |  |
|---|---|---|---|---|---|---|---|---|---|
| Level 1 | $ - | $ - | $ - | $ - | $ - | $ - | $ - | $ - |  |
| Level 2 | $ - | $ 25 | $ - | $ 10 | $ - | $ - | $ - | $ - |  |
| Level 3 | $ - | $ - | $ 50 | $ - | $ 75 | $ 40 | $ - | $ - |  |
| Level 4 | $ - | $ - | $ - | $ - | $ - | $ - | $ - | $ - |  |
| Level 5 | $ - | $ - | $ - | $ - | $ - | $ - | $ - | $ 550 | Unit Price, p |
| sum | $ - | $ 25 | $ 50 | $ 10 | $ 75 | $ 40 | $ 0 | $ 550 | $ 750 |

**Calculated** — Buyer Utilities for Selection Produced & Sold. $u = X^{[k,l]} \circ U^{[k,l]}$

|  | Attribute 1 | Attribute 2 | Attribute 3 | Attribute 4 | Attribute 5 | Attribute 6 | Attribute 7 | Attribute 8 |  |
|---|---|---|---|---|---|---|---|---|---|
| Level 1 | 0 | 0 | 0 | 0 | 0 | 0 | 0 | 0 |  |
| Level 2 | 71 | 69 | 0 | 45 | 0 | 0 | 23 | 0 |  |
| Level 3 | 0 | 0 | 46 | 0 | 65 | 15 | 0 | 0 |  |
| Level 4 | 0 | 0 | 0 | 0 | 0 | 0 | 0 | 0 |  |
| Level 5 | 0 | 0 | 0 | 0 | 0 | 0 | 0 | 1 | Total Utility, u |
| sum | 71 | 69 | 46 | 45 | 65 | 15 | 23 | 1 | 335 |

**Figure G.17** Supporting calculations for essential solver spreadsheets.

plot was determined by first calculating the profit for every design solution. For an eight-attribute, five-level product description matrix, there are over 16,000 possibilities. Based on the number of meaningful terms in the wrist videophone exercise, there are approximately 9000 possible solutions. Many times, different designs will have the same cost or would be priced the same as another. Therefore, the choice that gives the highest profit at a point is plotted. The contour plot of Fig. G.18 is based on fewer than 50 points, with coarse curve fitting in between. Although it looks smooth, the surface is actually like a bumpy mountain, hence the local maxima. When there is no solution from the solver, such as when the price (baseline plus premiums) is less than the range plotted, the profit is set to zero. In the wrist videophone, this occurs in the high-price, low-cost region. From this floor, the profit profile increases steeply to a plateau and then further to a high point at $205 cost, $750 price. The characteristics of the wrist videophone exercise are such that a very high price can still provide a moderate volume and, for the data provided, yields the maximum profit of $420 million.

It is important in these cases to actually choose the design that is closest to the market sweet spot and not an optimum point based on a high price. Although achievable, a high-priced product with high margins would attract competitors who will likely undercut your price and rapidly

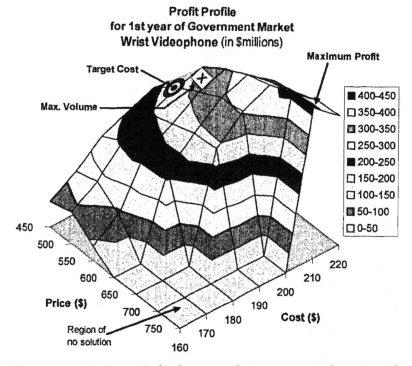

**Figure G.18** Profit profile for first year of government market wrist videophone with region of maximum profit highlighted.

erode your market share. The steep slope of the profit profile between costs of $200 and $210 illustrates the precariousness of this high-price solution. It is better to pick a more robust solution at a competitive price. A more robust solution occurs in the plateau, near the middle. You will gain insights by running the solver several times and choosing different objectives. Besides profit, you should determine other optimums. You should study the difference between the solution that maximizes profit, the one that minimizes the difference from the target cost, and the one that maximizes market share. All of these are calculated in the spreadsheet. Rearrange the profit objective function by choosing one of the other cells as your objective. Some of the results may be surprising. For example, the most profitable solution may not occur for the solution that would achieve the largest market share, as illustrated in Fig. G.19.

In this case, the maximum profit solution exceeds the target cost by 5% and the target price by 50%! Although the model does not include the impact of competitors, the high target price gap in this solution clearly indicates a precarious solution. The solution that maximizes market effectiveness or meets the target cost and price is more likely to succeed in the marketplace. In this case, these two moderately priced solutions differ

## Optimizing Results

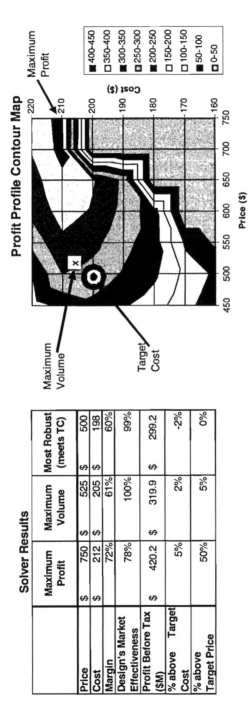

**Figure G.19** Various solutions to wrist videophone exercise using government market segment buyer utilities and costs. Solutions are shown on a profit profile contour map.

**Product Choices, X $^{[k, l]}$ for Maximum-Volume Solution**

|  | Attribute 1 | Attribute 2 | Attribute 3 | Attribute 4 | Attribute 5 | Attribute 6 | Attribute 7 | Attribute 8 |
|---|---|---|---|---|---|---|---|---|
|  | Color | Display Size | Video Speed | Stored Numbers | Talk Time | Messaging | Delivery | Base Price |
| Level 1 | 0 | 0 | 0 | 0 | 0 | 1 | 0 | 0 |
| Level 2 | 1 | 1 | 0 | 1 | 1 | 0 | 1 | 1 |
| Level 3 | 0 | 0 | 1 | 0 | 0 | 0 | 0 | 0 |
| Level 4 | 0 | 0 | 0 | 0 | 0 | 0 | 0 | 0 |
| Level 5 | 0 | 0 | 0 | 0 | 0 | 0 | 0 | 0 |
| sum | 1 | 1 | 1 | 1 | 1 | 1 | 1 | 1 |

**Product Choices, X $^{[k, l]}$ for Most Robust Solution**

|  | Attribute 1 | Attribute 2 | Attribute 3 | Attribute 4 | Attribute 5 | Attribute 6 | Attribute 7 | Attribute 8 |
|---|---|---|---|---|---|---|---|---|
|  | Color | Display Size | Video Speed | Stored Numbers | Talk Time | Messaging | Delivery | Base Price |
| Level 1 | 0 | 0 | 0 | 0 | 0 | 1 | 0 | 0 |
| Level 2 | 1 | 1 | 1 | 1 | 1 | 0 | 1 | 1 |
| Level 3 | 0 | 0 | 0 | 0 | 0 | 0 | 0 | 0 |
| Level 4 | 0 | 0 | 0 | 0 | 0 | 0 | 0 | 0 |
| Level 5 | 0 | 0 | 0 | 0 | 0 | 0 | 0 | 0 |
| sum | 1 | 1 | 1 | 1 | 1 | 1 | 1 | 1 |

**Figure G.20** Maximum volume (or market share) and most robust solution for the government market data in wrist videophone exercise. The only difference between these two is in video speed. The shaded boxes highlight the maximum profit solution when it is different.

in only one choice—video speed—as shown in Fig. G.20. Either one will do better than the high-priced solution.

Some readers may wonder whether the design suggested using the solver method gives the same solution when iterating with the model based on Sawtooth Software's ACA® simulator. The answer is, it does if the ACA® simulator has the same market data and buyer utility matrix. There are two differences between the buyer utility matrix used to derive the market share estimates of appendix F and the one used to illustrate the solver method. First, the one used to derive the market share contains data and thus influences from customers in all three market segments, not just government. Secondly, it includes one additional factor not included in the buyer utility matrix used in the solver method. In actual problems, an additional factor is used to compensate for attributes not included in the survey. This additional factor helps align results with actual market shares. When external factors, as this additional factor is often called, is turned off and the same market segments are used, the results agree. That said, the designs suggested by the solver approach give good results in the simulation.

In this illustration, buyer utility market data for only *one market segment* and only *one year* was used. To fully appreciate this approach, you will want to run the solver for other markets and the other years. In the wrist videophone example, as in most cases, the end result depends heavily on the market strategy selected. Design choices that focus on the markets that develop later, which in the wrist videophone case were the business and consumer markets, yield lower total profit results.

# Optimizing Results

## CONCLUSION

Using an iterative solver in the way presented here provides a framework for guiding teams to the design that maximizes profit or optimizes some other design variable. When the optimum profit solution is not the most robust in the market, this approach provides insights by showing the sensitivity of the design choices and pricing decisions. In actual problems, this approach helps the design team focus resources on the best solution and eliminates the need to invest in multiple, alternative approaches that will not meet the profit requirements.

> Congratulations! You have finished the Exercise.
> Please return to Chapter 7, page 101.

## GLOSSARY OF TERMS IN SOLVER SPREADSHEETS

| | |
|---|---|
| attribute | A product characteristic (e.g. color) |
| $\mathbf{B}^{[k,n]}$ | Best subsystem cost matrix. It contains best subsystem costs for each level as determined from the component costs. |
| $\mathbf{C}^{[k,n]}$ | Attribute cost matrix. It contains attribute costs for each level based on the best cost matrix, $\mathbf{B}$, and the subsystem mapping matrix, S. |
| $c$ | Product's unit cost |
| $e$ | Corresponding total product expense for the quantity produced |
| $\mathbf{E}^{[k,n]}$ | Attribute expense matrix. It contains the corresponding attribute expenses for each subsystem based on the subsystem mapping matrix, S. |
| $\mathbf{F}^{[n,m]}$ | Function–subsystem mapping matrix. It provides the fraction of subsystem contribution to each attribute. |
| $f_{ij}$ | The value of the $\mathbf{F}^{[m,n]}$ matrix of dimension $m \times n$, at location row $i$ and column $j$ |
| $f$ | Forecast |
| $k$ | Maximum number of attribute levels |
| $l$ | Total number of attributes (including base price) |
| $m$ | Number of attributes without price (e.g., $m = l - 1$) |
| $n$ | Total number of subsystems |
| level | A specific value of a product characteristic (e.g., black) |
| $P$ | Total profit |
| $p$ | Product's unit price |
| $\mathbf{P}^{[k,l]}$ | Table of prices expressed as base price and incremental prices for each attribute level |

| | |
|---|---|
| $\mathbf{S}^{[n,m]}$ | The subsystem mapping matrix is a variation of the attribute mapping matrix. It provides the fraction of each attribute allocated to each subsystem. |
| $s_{ij}$ | The value of the $\mathbf{S}$ matrix at location row $i$ and column $j$ |
| $t_i$ | Sum of terms in the $i$th row of the given matrix |
| $\mathbf{U}^{[k,l]}$ | Conjoint Analysis buyer utility matrix |
| $u$ | Total utility for a given design |
| $u_j|_{max}$ | Maximum buyer utility for the attribute $j$ |
| $v$ | Volume or quantity sold |
| $\mathbf{X}^{[k,l]}$ | Design and price choices (unknown in the beginning) |
| $\eta$ | Market effectiveness or efficiency of the design in the marketplace |

# Appendix

Sample Exercise

We have provided this appendix to illustrate the progress through the Exercise. This information may help make clear the types of information that should be generated, and the appearance of the results. If the reader is having difficulty with a point or two, seeing how one person handled the situation may be helpful. We have taken the inputs and results of an interested student of Target Costing who read each chapter of the draft manuscript and worked through each module of the Exercise. She is an individual without a technical or financial background and she got a good result, which helps reinforce the concept that Target Costing does not require professional technical or financial expertise.

The pages that follow contain reproductions of the inputs and worksheets that she generated, as well as the outputs at the end of the Exercise. The paragraphs that follow provide comments about each step.

**MODULE A: CREATE A BUSINESS WITH A STRATEGY**

(See Table A.1) The student named her company Best VideoPhone, Inc., or "BVP." She purchased the DataSearch forecast and the percentage of sales by segment. For the competitive information, she looked at the strengths, weakness and strategy, and the reverse engineering. Based on that, her initial strategy was to try to hold Government sales, and also try to capture a lot of the Consumer sales. It could be mentioned here that as the Exercise unfolded, later on she changed her strategy slightly to focus more strongly on the Government sector.

**TABLE A.1** Output for Module A—Create a Company with a Strategy

---

**Your Company's Name:**   Best VideoPhone, Inc (BVP)

**Your Strategy:**

Intended sales of the <u>new</u> product into each segment:

  Gov't.: __50__ %   Bus.: __10__ %   Cnsm.: __40__ %   Total=100%

How this will be done (3-4 sentences):

  Retain Government sales.

  New product will appeal to Gov't., Bus., and Consum.

  Reduce costs and have a competitive price.

Information that supports this strategy (2-4 sentences):

  Market-share data.

  Reverse-engineering information.

**Information Purchased in Module A** (fill in the cost for the ones that you purchased):

| Item Letter (A, B, C, ....) | Cost ($) |
|---|---|
| A | $ 50K |
| B | ----- |
| C | ----- |
| D | ----- |
| E | $ 100K |
| F | $ 50K |
| G | $ 200K |
| H | ----- |
| I | ----- |
| Total = | $ 400K |

---

## MODULE B: QUANTIFY CUSTOMER'S NEEDS

The student purchased the required worksheet (see page 243), as well as the market-feature table for all customers, the consultant's view of attributes and levels, and the Conjoint Analysis study. She also decided to focus almost completely on the Government sector, which was a shift from her original strategy. Based on the information she purchased, and on the utilities that pertain to the Government sector, she determined

# Sample Exercise

the relative importance of the attributes (see Table B.1). (Readers who are trying to address multiple sectors need to consider how utilities for different attributes and levels for different sectors need to be combined.) In this student's case, she determined that color, talk time, and display were the functional needs that required the most attention. Based on her analysis of the available information, she made her initial selection of "levels" for each of the attributes (Table B.2). This is a middle-of-the road wrist videophone, with a good amount of capability, but it isn't a top-of-the-line product.

TABLE B.1  Output for Module B (Part 1)—Quantify Customer's Needs

Your Company's Name: __Best VideoPhone, Inc (BVP)__

Relative Importance of Key Features or Attributes (for the New Product in its Marketplace):
(NOTE: Price is excluded.)

| Functional Need | Relative Importance (%) | (for Government) Preferred: |
|---|---|---|
| Color | 21% | Black |
| Display | 18% | 1-1/2" color |
| Speed | 14% | 24 f/sec (or even 12 f/sec) |
| Stored Numbers | 14% | 16 numbers |
| Talk Time | 21% | 16 hrs. standby, 30 min. talk |
| Message Service | 5% | None |
| Delivery | 7% | 72 hrs. |
| Total = | 100% | |

The Top 3 Attributes:
1. Color
2. Talk Time
3. Display

Information Purchased in Module B (fill in the cost for the ones that you purchased):

| Item Letter (A, B, C, ....) | Cost ($) | |
|---|---|---|
| A | $50K | (REQUIRED) |
| B | — | |
| C | — | |
| D | — | |
| E | $ 100K | |
| F | $ 100K | |
| G | $ 250K | |
| Total = | $ 450K | |

**TABLE B.2** Output for Module B (Part 2)—Select Product Features

Your Company's Name: _____Best VideoPhone, Inc (BVP)_____

**Features to be Provided in Your Company's New Product:**
(NOTE: These may not be your final selections. You could decide to change one or two of them later, when you are in Module E.)

| Attribute, and Level | Circle one number in each group | |
|---|---|---|
| **Color:** | | |
| Chrome | 1 | |
| Black (standard) | (2) | |
| Custom: brown, blue, other | 3 | |
| Gold | 4 | |
| **Display:** | | |
| 1.5 inch diagonal, black & white | 1 | |
| 1.5 inch diagonal, color | (2) | |
| 2.0 inch diagonal, color | 3 | |
| **Video Speed:** | | |
| industry standard (2 frames/second) | 1 | |
| standard video phone (12 frames/second) | 2 | ← 2nd choice |
| computer movie quality (24 frames/second) | (3) | ← 1st choice |
| **Stored Numbers:** | | |
| 4 numbers | 1 | |
| 16 numbers | (2) | |
| 32 numbers | 3 | |
| **Talk Time:** | | |
| 15-minute talk time, 10-hr. standby | 1 | |
| 30-minute talk time, 16-hr. standby | (2) | ← 1st choice |
| 60-minute talk time, 24-hr. standby | 3 | ← 2nd choice |
| **Messaging Service:** | | |
| None | (1) | |
| stores 2 minutes of voice messages | 2 | |
| stores 5 minutes of voice messages | 3 | |
| **Delivery:** | | |
| within 24 hours | 1 | |
| within 72 hours | (2) | |

## Sample Exercise

### Item A  Customer Preferences Worksheet
(All Attributes and Levels)

- Fill in buyer utilities relevant to your product strategy.
- Decide how you want to weight these attributes in your product.

Module B
Info. Item A
(REQUIRED)

Yes, I looked at this Item

| Wrist Videophone | Buyer Utilities | Max-Min | Relative Weight | Rel. Wt. (w/o Price) |
|---|---|---|---|---|
| **Color:** | | | 17% | 21% |
| Chrome ($5 discount) | 8 | | | |
| Black (standard) | 71 | 70 | | |
| Custom color exterior: brown, blue or other for $25 | 12 | | | |
| Gold case for $35 | 1 | | | |
| **Display:** | | | 15% | 18% |
| 1.5 inch diagonal black & white | 9 | | | |
| 1.5 inch color for $25 | 69 | 60 | | |
| 2.0 inch color for $50 | 42 | | | |
| **Video Speed:** | | | 11% | 14% |
| Industry standard: 2 frames/sec. ($25 discount) | 0 | | | |
| Standard video phone: 12 frames/sec. | 42 | 46 | | |
| Computer movie quality: 24 frames/sec. For $50 | 46 | | | |
| **Stored Numbers:** | | | 11% | 14% |
| 4 | 0 | | | |
| 16 for $10 | 45 | 45 | | |
| 32 for $25 | 15 | | | |
| **Talk Time:** | | | 17% | 21% |
| 10 hr. standby, 15 minute talk time | 0 | | | |
| 24 hr. standby, 60 minute talk time for $75 | 70 | 70 | | |
| 17 hr. standby, 30 minute talk time for $40 | 65 | | | |
| **Messaging Service:** | | | 4% | 5% |
| None | 33 | | | |
| Store 2 minutes of messages for $25 | 17 | 18 | | |
| Store 5 minutes of messages for $40 | 15 | | | |
| **Delivery:** | | | 6% | 7% |
| 24 hours for shipping premium | 0 | 23 | | |
| 72 hours | 23 | | | |
| **Selling Price:** | | | 18% | |
| $150 | 65 | | | |
| $250 | 72 | | | (Price not included) |
| $350 | 48 | 71 | | |
| $450 | 32 | | | |
| $550 | 1 | | | |
| Total: | | 403 | 100% | 100% |

## MODULE C: DETERMINE TARGET PRICE AND COST (PRODUCT LEVEL)

By this point the student realized that the cost of the offered information was a very small fraction of the company's annual revenues, and decided that it was best to err on the side of purchasing too much instead of too little. In this module she bought everything available. She triangulated on the experience curves that seemed to be most relevant to wrist videophones, and settled on an "80% curve"—that is, prices decline for every doubling of cumulative industry volume. (This was a little more aggressive than the 85% suggested in Item F, the consultant's view of the Government market.) She used the projected industry volumes from the DataSearch market forecast, which she had purchased as Item A in Module A. Starting with the 2009 cumulative volume of 500K units, she determined that the volumes in years 2010, 2011, 2012, and 2013 were 1000K, 2000K, 4000K and 7000K respectively. Beginning with a price of $525 in 2010 (1000K cumulative units), she projected an 80% curve: prices of $420, $336, and $290 as the volume successively doubles. This is the top line in Table C.1. The target prices of $420 and $336 happen to coincide with the cumulative volumes for the years 2011 and 2012; the $290 for 2013 was interpolated by eyeball. Finally, the student decided to attempt a 58% gross margin in all years; this led to target costs of $220, $177, $141, and $121 in the years 2010–2013.

# Sample Exercise

**TABLE C.1** Output for Module C—Determine Target Price and Cost (Product Level)

Your Company's Name: _____ Best VideoPhone, Inc. (BVP) _____

Experience Curves:

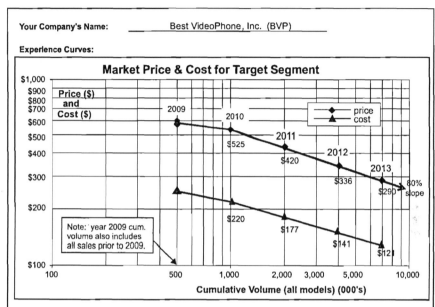

Target Prices, Gross Margins, and Costs (COGS):

| Year | Target Price | Target Margin | Target Cost |
|---|---|---|---|
| 2011 | $ 420 | 58 % | $ 177 |
| 2012 | $ 336 | 58 % | $ 141 |
| 2013 | $ 290 | 58 % | $ 121 |

Information Purchased in Module C (fill in the cost for the ones that you purchased)

| Item Letter (A, B, C, ....) | Cost ($) | |
|---|---|---|
| A | $50K | (REQUIRED) |
| B | $50K | |
| C | $50K | |
| D | $50K | |
| E | $50K | |
| F | $50K | |
| Total = | $300K | |

## MODULE D: DETERMINE COST TARGETS (SUB-SYSTEM LEVEL)

The output of this module is in Table D.1, and the value analysis worksheet that led to this output is on the following page. This worksheet was a required purchase, and the student supplemented it with the comprehensive Item I, which included virtually all the available information. The percentage contributions of all the subsystems to each of the functional needs are provided as part of the worksheet, so the student only had to enter the relative importance of features (from Module B), the target costs (from Module C), and the 2010 baseline costs for the subsystems (purchased in this module). While she used a spreadsheet to do the actual calculations, they can be done relatively quickly by hand with a pocket calculator. The student plotted the baseline 2010 costs as a function of the 2011 cost targets, and quickly saw that the biggest cost challenges were the camera, the telephone, and the controller subsystems.

# Sample Exercise

**TABLE D.1** Output for Module D—Determine Cost Targets (Subsystem Level)

Your Company's Name: _____Best VideoPhone, Inc. (BVP)_____

Cost Targets for Subsystems:

| Subsystem | 2011 Target | 2012 Target | 2013 Target | 2010 Baseline |
|---|---|---|---|---|
| Display | $ 54.16 | $ 43.15 | $ 37.03 | $ 40.00 |
| Controller | $ 40.71 | $ 32.43 | $ 27.83 | $ 65.00 |
| Camera | $ 9.74 | $ 7.76 | $ 6.66 | $ 35.00 |
| Telephone | $ 23.36 | $ 18.61 | $ 15.97 | $ 45.00 |
| Power | $ 23.81 | $ 18.96 | $ 16.27 | $ 20.00 |
| Housing & Assembly | $ 25.22 | $ 20.09 | $ 17.24 | $ 15.00 |
| Checksum | $ 177.00 | $ 141.00 | $ 121.00 | $ 220.00 |

2010 Baseline Cost vs. Cost Targets for year 2011:

**Information Purchased in Module D**
(fill in the cost for the ones that you purchased):

| Item Letter (A, B, C, ....) | Cost ($) | |
|---|---|---|
| A | $50K | (REQUIRED) |
| B | — | |
| C | — | |
| D | — | |
| E | — | |
| F | — | |
| G | — | |
| H | — | |
| I | $100K | |
| Total = | $150K | |

The 3 Areas Needing the Most Improvement:

1. _____Camera_____
2. _____Telephone_____
3. _____Controller_____

**Item A  Feature Subsystem Matrix and
Worksheet for Value Analysis.**

Module D
Info. Item A

- Put data in the shaded cells:
    The relative importance of the features to your customers.
    The target costs for *your company's* new product in 2011, 2012, and 2013.
- Calculate the value-based cost of each feature in 2011, 2012, and 2013.
- Calculate the resulting cost of each subsystem in *your company's* new product in 2011, 2012, and 2013.
- Put data in the shaded cells:
    The subsystem costs for *your company's* current product in 2010.

Yes, I looked at this Item

(*Note*: If you do the calculations for a single year, the results for the other years will simply scale with the target costs for those years. That is because we are assuming that the relative importance of each feature remains constant over the three-year period.)

## Sample Exercise

|  | Customers' Functional Needs |  |  |  |  |  | Allowable Costs, by Subsystem |  |  |  |
|---|---|---|---|---|---|---|---|---|---|---|
| Subsystem | Color | Display | Video Speed | Stored Numbers | Talk Time | Messages | Delivery | 2011 Target | 2012 Target | 2013 Target | 2010 Baseline |
| Customer Importance Weighting → | 21% | 18% | 14% | 14% | 21% | 5% | 7% | ← Data In | | | ↙ Data In |
| Value-Based Costs (2011) → | $ 37.17 | $ 31.86 | $ 24.78 | $ 24.78 | $ 37.17 | $ 8.85 | $ 12.39 | $ 177.00 | | | |
| Value-Based Costs (2012) → | $ 29.61 | $ 25.38 | $ 19.74 | $ 19.74 | $ 29.61 | $ 7.05 | $ 9.87 | | $ 141.00 | | |
| Value-Based Costs (2013) → | $ 25.41 | $ 21.78 | $ 16.94 | $ 16.94 | $ 25.41 | $ 6.05 | $ 8.47 | | | $ 121.00 | |
| | | | | | | | | | | | |
| Display | 60% | 65% | 25% | — | 10% | — | 10% | $ 54.16 | $ 43.15 | $ 37.03 | $ 40.00 |
| Controller | — | — | 50% | 50% | 30% | 40% | 10% | $ 40.71 | $ 32.43 | $ 27.83 | $ 65.00 |
| Camera | — | 15% | 15% | — | — | — | 10% | $ 9.74 | $ 7.76 | $ 6.66 | $ 35.00 |
| Telephone | — | — | 10% | 20% | 30% | 40% | 10% | $ 23.36 | $ 18.61 | $ 15.97 | $ 45.00 |
| Power | — | 20% | — | 30% | 20% | 15% | 10% | $ 23.81 | $ 18.96 | $ 16.27 | $ 20.00 |
| Housing & Assembly | 40% | — | — | — | 10% | 5% | 50% | $ 25.22 | $ 20.09 | $ 17.24 | $ 15.00 |
| Checksum | 100% | 100% | 100% | 100% | 100% | 100% | 100% | $ 177.00 | $ 141.00 | $ 121.00 | $ 220.00 |

## MODULE E: FIND PATHS TO THE TARGETS

The student soon discovered that, for most people, this is the most challenging module. Her output is in Table E.1, and her Product Cost Worksheet that led to this output is on the following page. Looking at the worksheet, she made her initial choices of components and used the year 2011 to work with. With her initial set of parts, she found out that seven of them were unsuitable—either they didn't provide the required functionality, or they were incompatible with other parts. She also got one pleasant surprise—one of the parts could serve a dual function. Also, the costs added up to be well above her target. The second iteration uncovered one more unsuitable part, but was closer in cost to the overall target. By the third iteration, she had a set of parts that were acceptable, and a cost

TABLE E.1  Output for Module E—Find Paths to the Targets

Your Company's Name: Best VideoPhone, Inc. (BVP)

**Final Set of Features to be Provided in Your Company's New Product:**

| Attribute, and Level | Circle one number in each group |
|---|---|
| **Color:** | |
| Chrome | 1 |
| Black (standard) | **2** |
| Custom: brown, blue, other | 3 |
| Gold | 4 |
| **Display:** | |
| 1.5 inch diagonal, black & white | 1 |
| 1.5 inch diagonal, color | **2** |
| 2.0 inch diagonal, color | 3 |
| **Video Speed:** | |
| industry standard (2 frames/second) | 1 |
| standard video phone (12 frames/second) | **2** |
| computer movie quality (24 frames/second) | 3 |
| **Stored Numbers:** | |
| 4 numbers | 1 |
| 16 numbers | **2** |
| 32 numbers | 3 |
| **Talk Time:** | |
| 15-minute talk time, 10-hr. standby | 1 |
| 30-minute talk time, 16-hr. standby | **2** |
| 60-minute talk time, 24-hr. standby | 3 |
| **Messaging Service:** | |
| None | **1** |
| stores 2 minutes of voice messages | 2 |
| stores 5 minutes of voice messages | 3 |
| **Delivery:** | |
| within 24 hours | 1 |
| within 72 hours | **2** |

**Final Costs, Margins, Prices:**

| New Product | Year 2011 | Year 2012 | Year 2013 |
|---|---|---|---|
| COGS (without discounts) | $ 182.25 | $ 157.00 | $ 137.75 |
| Volume Discount on Entire COGS<br>Number of outside suppliers = 2<br>If suppliers = 1, calculate 5% discount<br>If suppliers = 2, calculate 2% discount<br>If suppliers = 3 or 4, calculate 0% discount | $ 3.65 | $ 3.14 | $ 2.76 |
| Systems with only one outside supplier<br>If Controller, record undiscounted cost =<br>If Telephone, record undiscounted cost =<br>If Display, record undiscounted cost =<br>If Hsg. & Assy., record undiscounted cost =<br>Select most advantageous one (write name) | $<br>$ 35.75<br>$<br>$ | $<br>$ 31.50<br>$<br>$ | $<br>$ 27.00<br>$<br>$ |
| If Controller, or Telephone, calculate 20% discount<br>If Display, or Hsg. & Assy., calculate 10% discount<br>If none have single o/s supplier, use 0% discount | $ 7.15 | $ 6.30 | $ 5.40 |
| Net COGS (with discounts applied) | $ 171.46 | $ 147.56 | $ 129.60 |
| Final Choice of Gross Margins | 59% | 56% | 55% |
| Final Choice of Prices | $ 420.00 | $ 336.00 | $ 290.00 |
| Original TargetPrices = | $ 420.00 | $ 338.00 | $ 290.00 |
| Original Gross Margins = | 58% | 58% | 58% |

**Expenses:**

| | Amount |
|---|---|
| DEVELOPMENT (in 2011): | |
| Module A | $ 400 K |
| Module B | $ 450 K |
| Module C | $ 300 K |
| Module D | $ 150 K |
| Module E:<br>number of Quotations<br>looked at = 28<br>multiply by $20,000 | $ 560 K |
| Total Devel. Expense = | $ 1,860 K |
| YEARLY (in 2011,2012, 2013):<br>Sum of Yearly Expenses<br>for each Component = | $ 14.900 M |

## Sample Exercise

# PRODUCT COST WORKSHEET

| Sub-system/Component | Targets for Critical Yr. = 2011 | 1st Iteration Quot. # | 1st Iteration Supplier | 1st Iteration Cost ($) | 2nd Iteration Quot. # | 2nd Iteration Supplier | 2nd Iteration Cost ($) | 3rd Iteration Quot. # | 3rd Iteration Supplier | 3rd Iteration Cost ($) | Final Quot. # | Final Supplier | Year 2011 | Year 2012 | Year 2013 | Yearly Expenses ($M) |
|---|---|---|---|---|---|---|---|---|---|---|---|---|---|---|---|---|
| **Display** | $ 54.18 | | | $ 45.25 | | | $ 42.25 | | | $ 42.25 | | | $ 42.25 | $ 33.75 | $ 28.50 | |
| LCD | | 3 | A | $ 19.00 | 4 | B | $ 18.00 | | | ↑ | 4 | B | $ 16.00 | $ 11.00 | $ 9.00 | $ 1.700 |
| Driver Circuit | | 10 | A | $ 22.00 | | | ↑ | | | ↑ | 10 | A | $ 22.00 | $ 19.00 | $ 16.00 | $ 0.200 |
| Bezel | | 17 | B | $ 4.25 | | | ↑ | | | ↑ | 17 | B | $ 4.25 | $ 3.75 | $ 3.50 | $ 0.700 |
| **Controller** | $ 40.71 | | | $ 37.50 | | | $ 40.00 | | | $ 40.00 | | | $ 40.00 | $ 34.25 | $ 30.00 | |
| Processor | | 20 | In-House | $ 10.00 | 23 | B | $ 8.00 | | | ↑ | 23 | B | $ 8.00 | $ 7.00 | $ 6.50 | $ 0.450 |
| RAM | | 28 | B | $ 4.00 | | | ↑ | | | ↑ | 28 | B | $ 4.00 | $ 3.75 | $ 3.50 | $ 0.500 |
| FLASH | | 30 | A | $ 7.00 | | | ↑ | | | ↑ | 30 | A | $ 7.00 | $ 5.00 | $ 4.00 | $ 0.400 |
| ASIC | | 40 | C | $ 12.00 | 38 | B | $ 16.00 | | | ↑ | 38 | B | $ 16.00 | $ 14.00 | $ 12.00 | $ 0.800 |
| Timer Circuit | | 42 | A | $ 4.50 | 45 | A | $ 5.00 | | | ↑ | 45 | A | $ 5.00 | $ 4.50 | $ 4.00 | $ 0.800 |
| **Camera** | $ 9.74 | | | $ 42.00 | | | $ 30.00 | | | $ 32.00 | | | $ 32.00 | $ 29.00 | $ 26.00 | |
| Lens | | 47 | In-House | $ 12.00 | | | ↑ | | | ↑ | 47 | In-House | $ 12.00 | $ 11.00 | $ 10.00 | $ 0.750 |
| Image ASIC | | 52 | A | $ 30.00 | 54 | B | $ 18.00 | 55 | B | $ 20.00 | 55 | B | $ 20.00 | $ 18.00 | $ 16.00 | $ 0.950 |
| **Telephone** | $ 23.36 | | | $ 35.75 | | | $ 35.75 | | | $ 35.75 | | | $ 35.75 | $ 31.50 | $ 27.00 | |
| Speaker | | 57 | A | $ 5.00 | | | ↑ | | | ↑ | 57 | A | $ 5.00 | $ 4.75 | $ 4.50 | $ 0.550 |
| Microphone | | 62 | A | $ 2.75 | | | ↑ | | | ↑ | 62 | A | $ 2.75 | $ 2.75 | $ 2.50 | $ 0.550 |
| Transmitter | | 68 | B | (combo below) | (not needed) | | | (not needed) | | | (not needed) | | $ - | $ - | $ - | $ - |
| Receiver | | 72 | In-House | $ 18.00 | | | ↑ | | | ↑ | 72 | In-House | $ 18.00 | $ 15.00 | $ 12.00 | $ 1.750 |
| Power Amplifier | | 74 | A | $ 10.00 | | | ↑ | | | ↑ | 74 | A | $ 10.00 | $ 9.00 | $ 8.00 | $ 0.550 |
| **Power** | $ 23.81 | | | $ 25.00 | | | $ 18.00 | | | $ 18.00 | | | $ 18.00 | $ 15.00 | $ 14.00 | |
| Batteries | | 77 | A | $ 11.00 | 79 | B | $ 12.00 | | | ↑ | 79 | B | $ 12.00 | $ 10.00 | $ 9.00 | $ 1.000 |
| Rectifier Circuit | | 83 | A | $ 14.00 | 84 | B | $ 6.00 | | | ↑ | 84 | B | $ 6.00 | $ 5.00 | $ 5.00 | $ 0.750 |
| **Housing & Assembly** | $ 25.22 | | | $ 14.25 | | | $ 14.25 | | | $ 14.25 | | | $ 14.25 | $ 13.50 | $ 12.25 | |
| Housing & Strap | | 86 | A | $ 1.00 | | | ↑ | | | ↑ | 86 | A | $ 1.00 | $ 1.00 | $ 1.00 | $ 0.500 |
| Antenna | | 91 | A | $ 3.75 | | | ↑ | | | ↑ | 91 | A | $ 3.75 | $ 3.50 | $ 3.25 | $ 0.550 |
| Assembly | | 95 | In-House | $ 9.50 | | | ↑ | | | ↑ | 95 | In-House | $ 9.50 | $ 9.00 | $ 8.00 | $ 1.450 |
| **Total =** | $ 177.00 | | | $ 199.75 | | | $ 180.25 | | | $ 182.25 | | | $ 182.25 | $ 157.00 | $ 137.75 | $ 14.90 |

▓ = not acceptable or does not meet requirements

that was within about $5 of the target. She decided that she would get a discount by having only two outside suppliers, and another discount on the five-part telephone because she used only one outside supplier for it. This would bring her on target. She noted that she would be providing only 12 frames/sec video speed instead of the intended 24 frames/sec, but examination of the target market's buyer utilities (in Module B) indicated that there wasn't a big difference in preference. Her final set of product features is in Table E.1. This table also shows, in the lower left, her total product COGS in each of the years, and the discounts that applied. This led to net COGS of $171.46, $147.56, and $129.60 in 2011, 2012, and 2013. Comparing with her original targets of $177.00, $141.00, and $121.00, she saw that she hadn't quite met the 2012 and 2013 targets. She decided to keep her original market-based target prices and accept slightly smaller margins in the latter two years. She also recorded (lower right) all her "development expenses" (the information she purchased, including the 28 quotations), and the sum of the yearly expenses of the final set of selected components.

## MODULE F: GET FINANCIAL RESULTS

This student had access to the simulator software. When her data was entered and the marketplace simulation was run, she got results that added up to a 2010–2013 net income of $320.4 million. If we do the manual approximation of the simulation, her result was almost the same: $326.3 million. The manual simulation of the captured market share is shown on the Market Share Worksheet. She indicated the levels of the product attributes that she had selected, and she transcribed the appropriate "weighted utilities" into the "score" columns for the Government, Business, and Consumer markets. She then added the columns to get three subtotals (48.7, 26.4, and 16.0 for Government, Business, and Consumer.). She used the equations provided to calculate scores for the prices in 2011, 2012, and 2013, for all three market segments. Adding these to the previous subtotals, she got total "scores" for each product in each year for each market segment. These numbers led to a calculation of the market share of each segment that the new product would capture in each year (e.g., in 2011 this is 49.1% for Government, 29.6% for Business, and 9.9% for Consumer) Multiplying these percentages by the number of units known to be sold in each segment in each year (provided in the chart), the unit sales can be calculated. Adding them up over all segments and dividing by the total annual sales, she determined that her product would capture 36.0% of the total market in 2011, 37.3% in 2012, and 34.3% in 2013.

She put these percentages into the Financial Results Worksheet, along with her unit prices and COGS, her 2010 development expenses ($1,860 million), and her $14.9 million yearly expenses. When all the other elements were calculated according to the formulae provided, she discovered that in the three years 2011–2013 her old and new products

**Sample Exercise**

had captured a combined market share of about 50% and her return on sales was about 30%. These are good results and they help explain the relatively high four-year net income of $326.3 million.

We hope that this real-life example of one student's progress through the Exercise has helped the reader through any difficulties encountered.

# MARKET SHARE WORKSHEET

| Attribute and Level | Put a "1" next to the level you chose | GOVERNMENT Weighted Utility | GOVERNMENT Score (pick the wtd. Utility that corresponds to your choice) | BUSINESS Weighted Utility | BUSINESS Score (pick the wtd. Utility that corresponds to your choice) | CONSUMER Weighted Utility | CONSUMER Score (pick the wtd. Utility that corresponds to your choice) | Product's Market Share (units) |
|---|---|---|---|---|---|---|---|---|
| **Color:** | | | | | | | | |
| Chrome | | 1.4 | | 6.2 | | 2.7 | | |
| Black (standard) | 1 | 12.3 | 12.3 | 0.8 | 0.8 | 0.5 | 0.6 | |
| Custom: brown, blue, other | | 2.1 | | 2.8 | | 2.9 | | |
| Gold | | 0.2 | | 0.5 | | 1.0 | | |
| **Display:** | | | | | | | | |
| 1.5 inch diagonal, black & white | | 1.3 | | 2.9 | | 3.2 | | |
| 1.5 inch diagonal, color | 1 | 10.3 | 10.3 | 9.7 | 9.7 | 1.8 | 1.8 | |
| 2.0 inch diagonal, color | | 6.3 | | 1.9 | | 0.3 | | |
| **Video Speed:** | | | | | | | | |
| Industry standard (2 frames/second) | | 0.0 | | 7.3 | | 7.0 | | |
| standard video phone (12 frames/second) | 1 | 4.8 | 4.8 | 8.6 | 8.6 | 3.2 | 3.2 | |
| computer movie quality (24 frames/second) | | 5.3 | | 0.2 | | 0.0 | | |
| **Stored Numbers:** | | | | | | | | |
| 4 numbers | | 0.0 | | 1.6 | | 1.7 | | |
| 16 numbers | 1 | 5.0 | 5.0 | 3.0 | 3.0 | 0.9 | 0.9 | |
| 32 numbers | | 1.7 | | 1.8 | | 0.0 | | |
| **Talk Time:** | | | | | | | | |
| 15-minute talk time, 10-hr standby | | 0.0 | | 1.7 | | 6.5 | | |
| 30-minute talk time, 16-hr standby | 1 | 12.2 | 12.2 | 3.1 | 3.1 | 4.1 | 4.1 | |
| 60-minute talk time, 24-hr standby | | 11.3 | | 1.6 | | 0.2 | | |
| **Messaging Service:** | | | | | | | | |
| None | 1 | 1.5 | | 0.4 | | 4.7 | | |
| stores 2 minutes of voice messages | | 0.8 | 1.5 | 0.4 | 0.4 | 1.6 | 4.7 | |
| stores 5 minutes of voice messages | | 0.7 | | 0.3 | | 0.0 | | |
| **Delivery:** | | | | | | | | |
| within 24 hours | | 0.0 | | 0.2 | | 0.0 | | |
| within 72 hours | 2 | 1.3 | 2.6 | 0.4 | 0.4 | 0.4 | 0.8 | |
| **Score sub-total =** | | | 48.7 | | 26.4 | | 16.0 | |
| **Price** | | | | | | | | |
| $150 | | 11.5 | | 36.5 | | 73.2 | | |
| $250 | | 12.7 | | 30.3 | | 49.7 | | |
| $350 | | 8.5 | | 17.9 | | 16.2 | | |
| $450 | | 5.6 | | 9.8 | | 2.8 | | |
| $550 | | 0.2 | | 1.4 | | 0.0 | | |
| Your Price in 2011 = | $ 420.00 | | 4.5 | | 11.7 | | 7.0 | |
| Your Price in 2012 = | $ 336.00 | | 6.2 | | 15.4 | | 20.0 | |
| Your Price in 2013 = | $ 290.00 | | 8.4 | | 15.0 | | 21.0 | |
| | | Share of Segment | Total Score | Share of Segment | Total Score | Share of Segment | Total Score | |
| Total Score for 2011 | | 49.1% | 53.1 | 29.6% | 38.1 | 9.9% | 23.0 | |
| Total Score for 2012 | | 51.4% | 54.9 | 34.4% | 41.8 | 26.8% | 36.0 | |
| Total Score for 2013 | | 51.7% | 55.1 | 33.8% | 41.4 | 28.1% | 37.0 | |
| | | Total Market Segment | Sales in Segment | Total Market Segment | Sales in Segment | Total Market Segment | Sales in Segment | Total Market (units) | Total Units Sold | Units Market Share |
| Unit Sales in 2011 | | 400,000 | 196,326 | 200,000 | 59,153 | 150,000 | 14,873 | 750,000 | 270,351 | 36.0% |
| Unit Sales in 2012 | | 500,000 | 256,873 | 600,000 | 206,382 | 500,000 | 133,750 | 1,600,000 | 597,105 | 37.3% |
| Unit Sales in 2013 | | 600,000 | 309,959 | 1,200,000 | 406,024 | 1,600,000 | 449,840 | 3,400,000 | 1,165,823 | 34.3% |

If any number calculates to be less than zero, enter zero.

# Sample Exercise

## FINANCIAL RESULTS WORKSHEET

| FINANCIAL RESULTS | Year 2010 Old Product = Total | Year 2011 Old Product | Year 2011 New Product | Year 2011 Total | Year 2012 Old Product | Year 2012 New Product | Year 2012 Total | Year 2013 Old Product | Year 2013 New Product | Year 2013 Total | 2010-2013 Grand Total Net Income |
|---|---|---|---|---|---|---|---|---|---|---|---|
| Total Market Size (units) | 425,000 | | | 750,000 | | | 1,600,000 | | | 3,400,000 | |
| Old Product's Share (%) | 37% | 10% | | 46% | 13% | | 51% | 15% | | 50% | |
| New Product's Share (%) | | | 36.0% | | | 37.3% | | | 34.3% | | |
| Unit PRICE | $ 525.00 | $ 475.00 | $ 420.00 | | $ 408.00 | $ 336.00 | | $ 343.00 | $ 290.00 | | |
| Unit COGS | $ 220.00 | $ 199.00 | $ 171.46 | | $ 182.00 | $ 147.56 | | $ 160.00 | $ 129.60 | | |
| Revenue ($M) | $ 102.0 | $ 35.2 | $ 113.4 | $ 148.6 | $ 87.3 | $ 200.5 | $ 287.9 | $ 179.0 | $ 338.2 | $ 517.2 | |
| COGS ($M) | $ 42.8 | $ 14.8 | $ 46.3 | $ 61.1 | $ 39.0 | $ 88.1 | $ 127.0 | $ 83.5 | $ 151.1 | $ 234.6 | |
| Gross Margin ($M) | $ 59.2 | | | $ 87.6 | | | $ 160.8 | | | $ 282.6 | |
| Gross Margin (%) | 58% | | | 59% | | | 56% | | | 55% | |
| Direct Expenses ($M) (Incl. 2010 Devel. Expenses =) | $ 25.9 $ 1,860 | $ 21.0 | $ 14.900 | $ 35.9 | $ 41.1 | $ 14.900 | $ 56.0 | $ 72.5 | $ 14.900 | $ 87.4 | |
| Allocated Expenses ($M) | $ 7.4 | | | $ 10.3 | | | $ 16.0 | | | $ 25.0 | |
| NET INCOME ($M) | $ 25.9 | | | $ 41.4 | | | $ 88.8 | | | $ 170.2 | $ 326.3 |
| ROS % | 25% | | | 28% | | | 31% | | | 33% | Net Income |

# Glossary

The following terms are defined in the context of this book:

**Accounts Payable.** Money owed by a corporation to its suppliers, usually for goods and services provided.

**Accounts Receivable.** Money owed to a corporation by its customers, usually for goods and services provided.

**Allowable Cost.** The cost of a subsystem suggested by a feature-subsystem matrix as a result of distributing the target cost among the subsystems according to the importance of the various features to the customers.

**Architecture.** The physical structure of a product, determined by a combination of the relationships among its functional subsystems and those among its features and attributes seen by the customers.

**ASIC.** Application-specific integrated circuit. An IC that is designed for one specific application.

**Attribute.** A product characteristic (e.g., color) or feature.

**Attribute Cost Matrix, C.** A matrix of component cost totals by attribute based on the best cost matrix, **B**, and the subsystem mapping matrix, **S**.

**Basic Features.** Features of a product that all customers in a market segment want and are willing to pay for.

**Best Subsystem Cost Matrix, B.** A matrix of component cost totals at the subsystem level based on more detailed material costs. Used in Module G, it contains best subsystem costs for each level as determined from the component costs.

**Brainstorming.** A process whereby a group of people generate a large number of ideas for solving a specific problem.

**Buyer Utility.** In Conjoint Analysis, a measure of the value to a customer of a level or choice of an attribute.

**Competitive Analysis.** Analysis of the business circumstances, market strategies, technical capabilities, and product offerings of a company's competitors.

**Conjoint Analysis.** A quantitative method used to measure customer preferences based on ranking product options or answering a series of tradeoff questions.

**Cost of Goods Sold (COGS).** The direct costs of a manufactured product. COGS usually includes materials, labor, and manufacturing overhead ("load").

**Cross-Functional Team.** A group of people from a variety of functional areas in a company, with a variety of professional and business skills, formed to work on a common issue.

**Debt.** Certain types of money owed by a corporation. This includes short-term debt such as accounts payable, and long-term debt such as loans to the corporation.

**Depreciation.** The amount of money that a corporation has to set aside, usually according to a multi-year schedule, to account for the fact that capital assets (e.g., buildings, equipment) have a limited useful life and decline steadily in value.

**Design for "X" (DFX).** A design process that optimizes the product to reduce cost, time, and effort in the design, manufacture, delivery, and retirement of a product. "X" can be any of several product life-cycle stages such as manufacture, test, installation, etc. "X" can also represent considerations such as safety, environmental compliance, etc.

**Design Matrix, X.** A matrix representing the design and price choices (unknown in the beginning) to be made for a proposed product.

**DFX.** *See* Design for "X."

**DRAM.** Dynamic random-access memory, a device for storing information, used in computers.

**DSP.** Digital signal processor, a device for performing complex analysis of analog signals. Examples include speech recognition and undersea submarine surveillance.

**Equity.** The portion of a corporation that is owned by the shareholders, after all other liabilities have been satisfied.

**Expense.** Any of the other costs incurred in a corporation beyond the material, labor, and load associated with the production of the product. These include Research and Development, Marketing and Sales, and General and Administrative.

**Experience Curve.** A representation of the phenomenon that prices and costs of goods and services decline as a function of the cumulative experience gained in providing them. Sometimes called "learning curve."

**Function–Subsystem Matrix.** An array that shows the relationships between a product's features or attributes desired by customers and the subsystems that comprise the product. The matrix can then be used to distribute the product's target cost—first among the features according to the relative importance of those features to the customers, then among the subsystems according to the extent to which they contribute to the features.

**Forecast.** An estimate of the quantity of a product to be produced or needed for a given market.

**Function–subsystem Matrix.** *See* Feature–Subsystem Matrix.

**General and Administrative.** Costs in a corporation, not associated with producing products, which are incurred in the general running of the business. Examples include office expenses, executive salaries.

**House of Quality.** A pictorial representation of QFD. So named because diagram has the shape of a house. *See* QFD

**IC.** Integrated circuit. An electronic device that integrates a large number of functional circuits into a semiconductor device, and which is in a package that can be assembled into a system.

**Importance Table.** A table that lists the major features or attributes of a product and indicates their relative importance to a set of customers, usually as percentages adding up to 100%.

**Inventory.** Goods and services owned by a company that have not been purchased by customers. Types of inventory include raw materials (RMI) not yet embodied in products, work in process (WIP) which is partially finished, and finished goods (FGI) not yet sold.

**Kaizen.** A Japanese word that captures the concept of relentless "continuous improvement."

**L&L.** Labor and load. Most frequently associated with the labor and overhead costs incurred in a manufacturing process.

**Learning Curve.** A special case of an experience curve for one product or service (*see* Experience Curve).

**Level.** In a Conjoint Analysis survey, the possibilities for an attribute (e.g., red and green or levels for the product attribute, color).

**Market Efficiency, $\eta$.** A measure of market effectiveness or efficiency based on the straight sum of buyer utilities for a design divided by the maximum used in Module G. *See* Market Share.

**Market-Feature Table.** A table for relating market segments and the features that they want—Basic, Step-Up, and Premium.

**Market Segments.** Different subsets of a company's overall set of customers, each with different wants, needs, and willingness to pay. They are often differentiated by region or industry.

**Market Share.** A company's percentage of total sales in a given market.

**Materials.** The physical items out of which a product is made. They may include raw materials such as silicon, petroleum or electronic components, and functioning subsystems such as power supplies or disk drives. Note that one company's "finished goods" might be the next company's "raw materials."

**OEM.** Original equipment manufacturer. Usually refers to manufacturers of parts or subsystems that are embedded in another company's products. Sometimes OEM's make an entire product that is then branded with the selling company's name.

**Pareto.** Vilfredo Pareto (1848–1932), an Italian sociologist and economist who worked in Switzerland. He formulated the concept that a small subset of factors usually has the greatest aggregate effect on a phenomenon. "Pareto analysis" is involved with determining the most influential factors.

**Payables.** *See* Accounts Payable.

**Premium Features.** High-end or luxury features, usually ones with high profit margins, desired and willing to be paid for by a few customers in a market segment.

**Product Management.** The corporate function of managing the conception, introduction, marketing, and profitability of a product.

**Product Planning.** *See* Product Management.

**Profit.** The money left over after certain costs have been subtracted from revenues. There can be gross profit (after cost of goods sold), net operating profit before taxes, net operating profit after taxes, and so on.

**QFD.** Quality function deployment. A method of assessing customer value for functions and mapping them to subsystems.

**R&D.** Research and Development.

**Receivables.** *See* Accounts Receivable.

**Relative Importance Table.** *See* Importance Table.

**RF.** Radiofrequency, the portion of the electromagnetic spectrum that generally includes broadcast radio, television, radar, cellular telephony, microwave cooking, etc.

**RFMV.** Return to fair market value. A reserve account set up to cover the cost of obsolete inventory.

**ROM.** Read-only memory, a device for storing information, used in computers.

**ROS.** Return on sales. The net profit as a percentage of total sales.

**Royalties.** Money paid to a corporation, usually in consideration for allowing the use of its intellectual property such as patents and copyrights.

**Steering Team.** Target Costing steering team of management representatives from product management, market research, sales, competitive intelligence, research, product development, engineering, and manufacturing formed to oversee the project to achieve the target costs.

**Step-Up Features.** Features desired and willing to be paid for by some—but not all—customers in a market segment.

**Subsystem Mapping Matrix, S.** A variation of the function–subsystem mapping matrix. It provides the fraction of each attribute allocated to each subsystem (*see* Function–Subsystem Matrix).

**Target Cost.** The cost at which a product *must* be made to generate the target margin when sold at the target price.

**Target Margin.** The profit margin that must be achieved for the corporation to sustain itself as a healthy business. Most strongly influenced by the financial community and investors (owners).

**Target Price.** The market-based price at which a product can be sold. Most strongly influenced by customers and competitors.

**Unit Cost.** The cost of one instance of a mass-produced product.

**Unit Price.** The price of one instance of a mass-produced product.

**Utility.** *See* Buyer Utility.

**Value Engineering.** A discipline for determining costs of a product and its subsystems, based on value to the customer or user.

**Volume.** The quantity or number of units produced and sold over a given period of time.

**Warranty.** An obligation by a corporation to rectify or replace, at its own expense, products that become defective within a specified period of time.

# Index

Achieve the Target, 71–88, 108–109, 119, 125, 126, 133
Adaptive Conjoint Analysis (ACA), 34
Analysts' reports, 47
Attribute, 25, 221, 222

Balance sheet, 50–51
Basic features, 21, 59, 83, 90, 93
Best costs, 225–227
Bids and proposals, 46
Boeing 777 example, 19–20
Bottoms-up view, 46
Brainstorming, 75, 76, 79–81, 85, 105–107, 125
Bubble 76–77, 78
Buyer utility:
　calculating total utilities, 32
　definition, 25
　in case studies, 122

Can opener (*see* Electric can opener example)
Cellular base stations, 113–120
Champions, 104
Checklists, 109–111
Classifying Ideas, 81–82

COGS-Plus, 50
Comparable technologies, 46
Competitive analysis:
　description, 15
　competitors, 17
　customers, 18
Competitor knowledge, 17–18
Competitors' prices, 46
Conjoint Analysis:
　applying, 34–36
　in case studies, 120, 122
　caveats, 35–36
　description, 35
　determining buyer utilities, 26
　example, 25–34
　interpreting utilities, 28
　trade-off questions, 26
Constant dollars, 39
Continuous improvement, 6
Cost of goods sold (COGS), 4, 11, 48–49, 51, 89–94, 117
Cost plan, 93
Cost reduction, generic, 71, 75
Costs:
　full-stream, 48–51
　total, 10–11
　value-based, 58
　when set, 3

# Index

Cross-functional team(s), 36, 57–58, 76, 89–90, 104, 117, 120, 125, 126, 129
Customer knowledge, 18
Customer value matrix, 24–25

Define the Product, 15–38, 114, 120, 127, 134
Design for X (DFX), 78–79
Disruptive technologies, 47
DRAM, 42

Electric can opener example:
  buyer utilities, 29–30
  competitiveness, 23
  conjoint analysis for, 27–34
  cost targets for subsystems, 62–66
  customer value matrix, 25
  interpreting buyer utilities, 29–33
  mapping customer needs, 62–65
  market-feature table, 24
  subsystem experience curves, 67
  total addressable market, 25
  value of features, 61–62
Electronics cabinets example, 7–8, 113, 120–126
Exercise, discussion of, 13
Experience curves:
  cumulative experience, 89
  discussion, 41–45
  equation, 41
  history, 41
  steepness, 72–73
  use of, 128

Feature creep, 6, 83
Features:
  basic, 21
  step-up, 21
  premium, 21
Feature-subsystem matrix (*see* Funtion-subsystem matrix)

Financial results, 211–218, 253–255
Finding Paths, 75, 107–108, 191–209, 250–252
Functional Teams (*see* Cross-functional teams)
Function-subsystem matrix, 61–62

Gross margin, 11, 49
Gross profit, 49

Idea proposal form, 82
Income statement, 48–49

Jumbo telecom switch, 133–134

Kaizen, 4, 6

Learning curves (*see* Experience curves)
Level, 25, 221, 222

Maintain Competitive Costs, 89–100, 108–109
Margin, 51
Market research, 15
Market-adequate design, 59
Market-driven design, 77
Market-feature table, 20–22, 120, 126, 127, 133
Matrix:
  attribute cost, 223, 228
  attribute expense, 223, 228–229
  buyer utility, 221–224
  design, 223, 225
  operations, 223
  price, 221, 224–225
  subsystem mapping, 223, 226
Misunderstood wants and needs, 16
Moore's law, 42–43
Motivation, 2

Optical interface unit (OIU), 113, 126–133

Pareto analysis, 73, 76–78

# Index

Portfolio level reviews, 94–95
Premium features, 21, 83
Price:
  discussion, 39–41
  starting with, 5
  target, 45, 169–178, 244–245
  value-based, 58
Product configurations, 90
Product development cycle, 3–4
Product planning, 15
Product success and failure, 16
Profit:
  contour plots of, 232–235
  equation for, 221, 230
  optimization of, 221, 230–236
  results and observations, 231–236
  spreadsheet for optimization of, 222, 232–233
Public information, 47

Quantifying gaps, 95–98
Quantifying needs, 24–25, 155–167, 240–243

Readiness for target costing, 102–104
Readouts, 108
Reverse engineering, 46
Review meetings, 98–99
Robustness, 68–69
Roles of organizations, 91
Roles of target costing, 91

Sawtooth software, 34, 137
Set the Target:
  in case studies, 115, 125, 127
  product level, 39–54
  subsystem level, 55–70
Simulator, ACA, 219–220
Solver, Microsoft Excel function, 219–220, 231

Steering team, 74
Step-up features, 21, 83
Stockbrokers, 48
Subdividing the target, 55
Supplier target costing, 79–82

Target cost:
  calculating subsystem costs, 63–65
  equation, 5
  setting, 48, 169–178, 244–245
  subdividing, 55, 72–74
  subsystem, 67–68, 179–189, 244–249
  validating subsystem costs, 66–69
Target costing:
  benefits, 12
  definition, 1–2, 7, 10
  discipline, 10
  objectives, 11
  product families, 90–95
  steps, 4
  time and effort required, 105
Teams (*see* Cross-functional teams)
Telecommunications equipment example, 8–9
Templates:
  configuration, 92
  cost plan, 93
  ideas, 91, 93
  reviewing 90, 93, 98
Tools and techniques figure, 102
Total cost of ownership (TCO), 5

Utility (*see* Buyer utility)

Value analysis (*see* Value engineering)
Value-based prices and costs, 58–60
Value engineering:
  customer needs, 62
  description, 58–60

[Value engineering:]
  mapping customer needs to subsystems, 62–65
  value to the customer, 61
Value index, 65–66
Value matrix (*see* Customer value matrix)

Variables, 2

Willingness to pay, 5, 15, 21, 22, 47, 60
Wrist video phone:
  components of, 140
  scenario, 138–141